The Birth of
Numerical
Analysis

The Birth of Numerical Analysis

Editors

Adhemar Bultheel • Ronald Cools

Katholieke Universiteit Leuven, Belgium

World Scientific

NEW JERSEY · LONDON · SINGAPORE · BEIJING · SHANGHAI · HONG KONG · TAIPEI · CHENNAI

Published by

World Scientific Publishing Co. Pte. Ltd.

5 Toh Tuck Link, Singapore 596224

USA office: 27 Warren Street, Suite 401-402, Hackensack, NJ 07601

UK office: 57 Shelton Street, Covent Garden, London WC2H 9HE

British Library Cataloguing-in-Publication Data
A catalogue record for this book is available from the British Library.

THE BIRTH OF NUMERICAL ANALYSIS

ISBN-13 978-981-283-625-0
ISBN-10 981-283-625-X

Printed in Singapore.

Preface

1 The limitations of computers

In Wikipedia, numerical analysis is described as that part of mathematics where algorithms for problems of continuous mathematics are studied (as opposed to discrete mathematics). This means that it is especially dealing with real and complex variables, the solution of differential equations and other comparable problems that feature in physics and engineering. A real number has in principle an infinite number of digits, but on a digital computer, only a finite number of bits is reserved to store a (real) number. This memory restriction implies that only rounded, approximating values of only finitely many real numbers can be stored. The naive idea of the early days of digital computers was that they would not make the same "stupid errors" that human computers sometimes made, like transcription errors, reading errors, wrong signs, etc. This euphoria was however soon tempered when it was realized that computers in fact make errors in practically every calculation. Small errors indeed, but nevertheless a lot of errors. And all these small errors can accumulate and grow like a virus through the many elementary computations made which could eventually give a result that is quite different from the exact one.

2 A birthday?

A careful analysis of this propagation of errors when solving a linear system of equations was first published in a paper by John von Neumann and Herman Goldstine: *Numerical inverting of matrices of high order*, published in the November issue of the *Bulletin of the American Mathematical Society* in 1947. Because this was the first time that such an analysis was made, this paper is sometimes considered to be the start of modern numerical analysis. Of course numerical calculations were done long before that paper and problems from physics and engineering had been solved earlier, but the scale and the complexity of the computations increased drastically with the introduction of digital computers. The "large systems" to which the title of the paper refers, would not be called "large" at all by current standards. It is claimed in the paper that "serious problems can occur" if one wants to solve systems of more than ten equations. In a subsequent footnote, it is suggested that it would probably be possible in the future to solve systems of a hundred equations. If we know that the PageRank of Google can be computed by manipulating systems of approximately ten billion equations, then it should be clear we have come a long distance.

3 Sixty years young "back to the roots of the future"

If the publication of the von Neumann-Goldstine paper is indeed the start of numerical analysis, then November 2007 would be the moment that numerical analysis can celebrate its sixtieth birthday. This inspired the scientific research network *Advanced Numerical Methods for Mathematical Modeling*, a consortium of numerical analysis groups in Flanders, to organize a two-day symposium at the Department of Computer Science of the K.U.Leuven (Belgium) entitled *"The birth of numerical analysis"*. The idea of this symposium was to invite a number of speakers who were already active numerical analysts around the middle of the twentieth century or shortly after and hence were co-founders of the modern discipline. They came to witness about their experience during the early days and/or how their respective subdomains have evolved. Back to the roots is an important general cultural paradigm, and it is none the less true for numerical analysis. To build a sound future, one must have a thorough knowledge of the foundations. Eleven speakers came to Leuven on October 29-30, 2007 for this event to tell their personal story of the past and/or give their personal vision of the future. In the rest of this preface we give a short summary of their lectures. Most of them have also contributed to these proceedings.

4 Extrapolation

The start of the symposium was inspired by extrapolation. In many numerical methods, a sequence of successive approximations is constructed that hopefully converges to the desired solution. If it is expensive to compute a new approximation, it might be interesting to recombine a number of the last elements in the sequence to get a new approximation. This is called extrapolation (to the limit). In the talk of *Claude Brezinski*, a survey was given of the development of several extrapolation methods. For example for the computation of π, Christiaan Huygens used in the seventeenth century an extrapolation technique that was only rediscovered by Richardson in 1910, whose name has been attached to the method. Another extrapolation method was described by Werner Romberg in 1955. He tried to improve on the speed of convergence of approximations to an integral generated by the trapezium rule. This method is now called Romberg integration. After the introduction of digital computers, the improvements, generalizations, and variations of extrapolation techniques were numerous: Aitken, Steffensen, Takakazu, Shanks, Wynn, QD and epsilon algorithms are certainly familiar to most numerical analysts. Read more about this in Brezinski's contribution of this book.

> Claude Brezinski ($^\circ$1941) is emeritus professor at the University of Lille and is active on many related subjects among which extrapolation methods but also Padé approximation, continued fractions, orthogonal polynomials, numerical linear algebra and nonlinear equations. He has always shown a keen interest in the history of science, about which he has several books published.

The talk of *James Lyness* connected neatly with the previous talk. Romberg integration for one-dimensional integration also got applications in more-dimensional integration, but the first applications only came in 1975 for integrals over a simplex and integrands with a singularity. Nowadays this has become an elegant theory for integration of functions with an algebraic or logarithmic singularity in some vertices of a polyhedral domain of integration. The integration relies on three elements. Suppose we have a sequence of quadrature formulas $Q^{[m]}\{f\}$ that converge to the exact integral: $I\{f\} = Q^{[\infty]}\{f\}$. First one has to write an asymptotic expansion of the quadrature around $m = \infty$. For example $Q^{[m]}\{f\} = B_0 + B_1/m^2 + B_4/m^4 + \cdots + B_{2p}/m^{2p} + R_{2p}(m)$ with $B_0 = I\{f\}$. This is just an example and the form of the expansion should be designed in function of the singularity of the integrand. Next, this is evaluated for say n different values of m, which results in a system of n linear equations that is eventually solved for the B_k, in particular $B_0 = I\{f\}$. The moral of the story is that for further development of multidimensional extrapolation quadrature one only needs three simple elements: a routine to evaluate the integrand, a routine implementing the quadrature rule, and a solver for a linear system. Of course the most difficult and most creative part is to find the appropriate expansion.

The contribution of Lyness is included in these proceedings.

James Lyness (°1932) is employed at the Argonne National Laboratory and the University of New South Wales. His first publications appeared mainly in physics journals, but since he published his first paper on N-dimensional integration (co-authored by D. Mustard and J.M. Blatt) in *The Computer Journal* in 1963, he has been a leading authority in this domain with worldwide recognition.

5 Functional equations

The afternoon of the first day was devoted to functional equations.

From an historical point of view, we could say that the method of Euler for the solution of ordinary differential equations (1768) is the seed from which all the other methods were deduced. That is how it was presented in the lecture of *Gerhard Wanner*. Runge, Heun and Kutta published their methods around 1900. These were based on a Taylor series expansion of the solution and the idea was to approximate it in the best possible way. This means that for a small step h, the difference between the true and the approximating solutions was $O(h^p)$ with order p as high as possible. Such an approach quickly leads to very complicated systems of equations defining the parameters of the method. Therefore in the sixties and seventies of the previous century, much effort was put in a systematic treatment by, e.g., Butcher, Hairer and Wanner.

On the other hand, multistep methods are among the offspring of techniques by Adams and Bashforth which date back to 1885. These predict the next value of the solution using not only the last computed point, such as Runge-Kutta methods do, but they use several of the previously computed points to make the prediction. Dahlquist published in 1956 the generalized linear multistep methods. Important efforts have been made to improve the step control and the stability.

Gerhard Wanner (°1942) is professor at the University of Genève and ex-president of the Suisse Mathematical Society. He wrote together with Hairer several books on analysis and differential equations. The historical aspects always played an important role. He has had scientific contacts at all "levels" from 2 meter below sea level (the Runge-Kutta symposium at the CWI in Amsterdam on the occasion of 100 years of Runge-Kutta methods in 1995) to the top of the Mont Blanc at 4807 meters above sea level where he hiked together with Hairer.

The further development was picked up in the talk by *Rolf Jeltsch*. His main topic was the evolution of the concept of stability when solving stiff differential equations.

Stiff differential equations form a problem for numerical solution methods because the dynamics of the solution have components at quite different scales. Researchers wanted to design methods that computed numerically stable (bounded) solutions when the theoretical solution was supposed to be stable. Dahlquist proved in 1963 his well known second barrier for multistep methods. It stated that there was not an A-stable method of order higher than two. The A stands for absolute, which means that the numerical method computed a stable solution, whatever the step size is. This started a quest for other types of methods and gave rise to a whole alphabet of weaker types of stability.

Rolf Jeltsch (°1945) is professor at the ETH Zürich. He is a former president of the EMS (1999-2002) and of the Suisse Mathematical Society (2002-2003). Since 2005 he is president of the "Gesellschaft für Angewandte Mathematik und Mechanik" (GAMM). In the nineteen seventies his main research topic was ordinary differential equations. Since the nineteen eighties, he focussed more on hyperbolic partial differential equations and large scale computations in engineering applications.

Unfortunately, the contributions by Wanner and Jeltsch could not be included in these proceedings but the editors were happy to find a valuable replacement to cover the area of differential equations. *John Butcher* provided a text in which he reports on the contribution of New Zealanders, which includes his own, to numerical analysis in general and differential equations in particular. So he links up the European and New Zealand numerical scene. His personal reminiscences bring about a broader historical perspective.

John Butcher (°1933) is professor emeritus at the Department of Mathematics, of the University of Auckland. His main interests are in the numerical solution methods for ordinary differential equations. He is considered to be the world's leading specialist on Runge-Kutta methods. His 1987 book on the numerical solution of ordinary differential equations and their subsequent editions of 2003 and 2008, are considered the best reference work on this subject.

Herbert Keller was the dean of the company. He replaced Philip Davis, who first agreed to attend but eventually was not able to come to the meeting. The message of Keller was that singularities have always played an important role in numerical computations and that they were not always given the attention they deserve. This starts with such a simple thing as dividing by zero (or by something "almost zero"). This is obviously a fundamental issue. In Gaussian elimination for solving a linear system of equations, dividing by a small diagonal element may completely destroy the accuracy of the algorithm. But, it is as important to take care of a singularity of an integrand when designing good numerical quadrature formulas. This was already shown in the talk by Lyness. Another example is encountered when solving nonlinear equations where the Jacobian evolves during the iteration towards a matrix that is (almost) singular. This kind of difficulties certainly appears in more complex large scale problems connected with differential or integral equations, dynamical systems etc.

Herb Keller was in excellent shape during the symposium and full of travel plans. However, few months after the symposium, we received the sad news that Keller had passed away on January 26, 2008. So there is no contribution by him about his talk. Hinke Osinga, who had an interview with Keller published in the DSWeb Magazine, was kind enough to slightly adapt her text and this is included instead.

Herbert Keller ($^\circ$1925 – †2008) was emeritus professor at the California Institute of Technology. Together with E. Isaacson, he is the author of the legendary book "Analysis of Numerical Methods" that was published in 1966 by J. Wiley. His scientific contributions mainly dealt with boundary value problems and methods for the solution of bifurcation problems.

The first day was concluded by the lecture of *Kendall Atkinson* who spoke about his personal vision on the evolution in research related to the solution of integral equations. He emphasized the use of functional analysis and operator theory for the analysis of numerical methods to solve this kind of equations. The origin points to a paper by Kantorovich "Functional analysis and applied mathematics" that appeared in Russian in 1948. It deals among other things with the solution of Fredholm integral equations of the second kind. Atkinson summarizes the most important methods and the results that were obtained: degenerate kernel approximation techniques in which the kernel is written as $K(s,t) = \sum_i \alpha_i(s)\beta_i(t)$, projection methods (the well known Galerkin and collocation methods where the solution is written as a linear combination of basis functions with coefficients that are fixed by interpolation or by orthogonality conditions), and the Nyström method that is based on numerical integration. The details of his lecture can be found in these proceedings.

Kendall Atkinson ($^\circ$1940) is emeritus professor at the University of Iowa. He is an authority in the domain of integral equations. His research encompasses radiosity equations from computer graphics and multivariate approximation, interpolation and quadrature. He wrote several books on numerical analysis and integral equations.

6   The importance of software and the influence of hardware

The first lecture of the second day was given by *Brian Ford*. He sketched the start and the development of the NAG (Numerical Algorithms Group) software library. That library was the first collection of general routines that were not just focussing on one particular kind of numerical problem or on one particular application area. Also new was that it came out of the joint effort of several researchers coming from different groups. Ford started up his library in 1967, stimulated by his contacts with J. Wilkinson and L. Fox. Being generally applicable, well tested, and with good documentation it was an immediate success. The official start of the Algol and Fortran versions of the NAG library is May 13, 1970. The algorithms are chosen on the basis of stability, robustness, accuracy, adaptability and speed (the order is important). Ford then tells about the further development and the choices that had to be made while further expanding the library and how this required an interplay between numerical analysts and software designers. He concludes his talk with an appeal to young researchers to work on the challenge put forward by the new computer architectures where multicore hardware requires a completely new implementation of the numerical methods if one wants to optimally exploit the computing capacity to cut down on computer time and hence to solve larger problems.

The contribution of B. Ford is included in these proceedings. More on the (r)evolution concerning hardware and its influence on the design and implementation of numerical software can be found in the next contribution by J. Dongarra.

Brian Ford (°1940) is the founder of the NAG company and was director until his retirement in 2004. He received a honorary doctorate at the University of Bath and was given a OBE (Officer of British Empire) for his achievements. Under his leadership, NAG has developed into a respected company for the production of portable and robust software for numerical computations.

The conclusion of B. Ford was indeed the main theme in the lecture of *Jack Dongarra*. In his lecture he describes how, since about 1940, the development of numerical software, the hardware, and the informatics tools go hand in hand. Early software was developed on scalar architectures (EISPACK, LINPACK, BLAS '70) then came vector processors ('70-'80) and parallel algorithms, MIMD machines (ScaLAPACK '90), and later SMP, CMP, DMP, etc. The multicore processors are now a reality and a (r)evolution is emerging because we will have to deal with multicore architecture that will have hundreds and maybe thousands of cores. Using all this potential power in an efficient way by keeping all these processors busy, will require a total re-design of numerical software.

A short version of the lecture, concentrating on a shorter time-span, has been included in these proceedings.

Jack Dongarra (°1950) is professor at the University of Tennessee where he is the leader of the "Innovative Computing Laboratory". He is specialized in linear algebra software and more generally numerical software on parallel computers and other advanced architectures. In 2004 he received the "IEEE Sid Fernbach Award" for his work in HPC (High Performance Computing). He collaborated on the development of all the important software packages: EISPACK, LINPACK, BLAS, LAPACK, ScaLAPACK, etc.

7 Approximation and optimization

More software in the lecture of *Robert Plemmons*. The thread through his lecture is formed by nonnegativity conditions when solving all kinds of numerical problems. First a survey is given of historical methods for the solution of nonnegative least squares problem. There one wants to solve a linear least squares problem where the unknowns are all nonnegative. Also the factorization of two nonnegative matrices (NMF) was discussed. The latter is closely connected with data-analysis. Other techniques used here are SVD (singular value decomposition) and PCA (principal component analysis). These however do not take the nonnegativity into account. Around the nineteen nineties ICA (independent component analysis) was introduced for NMF. This can also be formulated as BSS (blind source separation). In that kind of application, a mixture of several sources is observed and the problem is to identify the separate sources. In more recent research, one tries to generalize NMF by replacing the matrices by tensors. There are many applications: filter out background noise from an acoustic signal, filtering of e-mails, data mining, detect sources of environment pollution, space research SOI (space object identification) etc.

Read more about this in the contribution by Chen and Plemmons in these proceedings.

Robert J. Plemmons (°1938) is Z. Smith Reynolds professor at the Wake Forest University, NC. His current research includes computational mathematics with applications in signal and image processing. For example, images that are out of focus are corrected, or atmospheric disturbances are removed. He published more than 150 papers and three books about this subject.

Michael Powell gave a survey of the successive variants of quasi-Newton methods or methods of variable metric for unconstrained nonlinear optimization. He talked about his contribution to their evolution of the prototypes that were designed in 1959. These methods were a considerable improvement over previous methods that were popular before that time which were dominated by conjugate gradients, pure Newton, and direct search methods. In these improved methods, an estimate for the matrix of second derivatives is updated so that it does not have to be recomputed in every iteration step. These methods converge fast and the underlying theory is quite different from the corresponding theory for a pure Newton method. Mike Powell then evoked methods derived from the previous ones that are important in view of methods for optimization problems with constraints. Further methods discussed are derivative free methods, and the proliferation of other types like simulated annealing, genetic algorithms etc. A summary of this presentation can be found in these proceedings.

Michael J.D. Powell (°1936) is professor at Cambridge University. He has been active in many domains. His name is for example attached to the DFP (Davidon-Fletcher-Powell) method (a quasi-Newton method for optimization), and a method with his name that is a variant of the Marquardt method for nonlinear least squares problems. But he is also well known for his work in approximation theory. The Powell-Sabin splines are still an active research area in connection with subdivision schemes for the hierarchical representation of geometric objects or in the solution of partial differential equations.

8   And some history

The closing lecture of the symposium was given by *Alistair Watson*. With some personal touch, he sketched the early evolution of numerical analysis in Scotland. In a broader sense, numerical analysis has existed for centuries of course. If you restrict it to numerical analysis as it was influenced by digital computers, then 1947 is a good choice to call that the start. But thinking of the more abstract connection between numerics and computers, then one should probably go back to 1913 when papers by Turing were published. That is where Watson starts his account of the history. More concretely, in Scotland, the start is associated with the work of Whittaker and later Aitken who were appointed in Edinburgh. In the "Mathematical Laboratory", founded in 1946, computations were done by means of pocket calculators. It was only in 1961 that this was coined to be a "Numerical Analysis" course. In that year Aitken claimed to have no need for a digital computer. It arrived anyway in 1963. Then things start to move quickly. More numerical centers came into existence in St Andrews and later in Dundee. The University of Dundee became only independent of St Andrews in 1967 and soon the gravity center of numerical analysis had moved to Dundee. From 1971 until 2007, the biennial conference on numerical analysis was held in Dundee. In the most recent years, the University of Strathclyde (Glasgow) seems to be the new attraction pole for numerical analysis in Scotland.

A longer write-up of this historical evolution is included in these proceedings.

The focus of the work of Alistair Watson (°1942) is numerical approximation and related matters. This can be theoretical aspects, but also elements from optimization and linear algebra. He is FRSE (Fellow of the Royal Society of Edinburgh) and he is probably best known by many for his involvement in the organization of the Dundee conferences on numerical analysis.

9 And there is more

Of the eleven lecturers at the symposium, the youngest was 57 and the oldest 82, with an average just over 68. All of them showed a lot of enthusiasm and made clear that whatever the exact age of numerical analysis, be it sixty years or a hundred years or even centuries, there still remains a lot to be done and the challenges of today are greater than ever.

Of course we would have liked to have invited many more people to lecture and cover more subjects at the symposium, but time and budget was finite. Some important people who had planned to come, finally decided for diverse reasons to decline. One of these was *Gene Golub* ($\circ$1932 – †2007), the undisputed godfather of numerical linear algebra, who, sad to say, passed away just after the symposium on November 19, 2007. Another name we would have placed on our list as a speaker for his contributions to numerical integration would have been *Philip Rabinowitz* ($\circ$1926 – †2006) had he still be among us. We are fortunate to have obtained permission of the American Mathematical Society to reprint an obituary with reminiscences of Phil Davis and Aviezri Fraenkel that was first published in the Notices of the AMS in December 2007.

Phil Davis also wrote up some personal reminiscences of what it was like in the very early days, when numerical analysis was just starting in the years before and during WW-II.

Philip Davis ($\circ$1923), currently emeritus professor at Brown University. He is well known for his work in numerical analysis and approximation theory, but with his many columns, and books, he also contributed a lot to the history and philosophy of mathematics. His books on quadrature (together with Ph. Rabinowitz) and on interpolation and approximation are classics. He also collaborated on the Abramowitz-Stegun project *Handbook of Mathematical Functions*. He started his career as a researcher in the Air Force in WW-II, and joined the National Bureau of Standards before going to Brown.

Finally, we found *Robert Piessens* kind enough to write another contribution for this book. His approach is again historical and sketches first the work of Chebyshev about linkage instruments, a mechanical tool to transform a rotation into a straight line. This is how Chebyshev polynomials came about. He continues by illustrating how the use of Chebyshev polynomials has influenced the research of his group at K.U.Leuven, since it turned out to be a powerful tool in developing methods for the numerical solution of several problems encompassing inversion of the Laplace transform, the computation of integral transforms, solution of integral equations and evaluation of integrals with a singularity. The latter resulted in the development of the QUADPACK package for numerical integration.

Robert Piessens ($\circ$1942) is emeritus professor at the Department of Computer Science at the K.U.Leuven, Belgium. He was among the first professors who started up the department. His PhD was about the numerical inversion of the Laplace transform. He is one of the developers of the QUADPACK package which has been a standard package for automatic numerical integration in one variable. Originally written in Fortran 77, it has been re-implemented in different environments. It is available via netlib, several of its routines have been re-coded and are integrated in Octave and Matlab, the Gnu Scientific Library (GSL) has a C-version, etc.

10 Acknowledgements

The symposium was sponsored by the FWO-Vlaanderen and the FNRS, the regional science foundations of Belgium. We greatefully acknowledge their support.

Leuven, March 2009   Adhemar Bultheel and Ronald Cools

From left to right:
H. Keller, C. Brezinski, G. Wanner, B. Ford, A. Watson, K. Atkinson,
R. Jeltsch, R. Plemmons, M. Powell, J. Lyness, J. Dongarra.
In front: the organizers:
R. Cools and A. Bultheel

G. Wanner, C. Brezinski, M. Powell   J. Dongarra, B. Ford, H. Keller

J. Lyness, R. Jeltsch   B. Plemmons, M. Powell

R. Jeltsch, A. Watson   B. Ford, K. Atkinson

Table of Contents

Preface v

Some pioneers of extrapolation methods 1
  Claude Brezinski

Very basic multidimensional extrapolation quadrature 23
  James N. Lyness

Numerical methods for ordinary differential equations: early days 35
  John C. Butcher

Interview with Herbert Bishop Keller 45
  Hinke M. Osinga

A personal perspective on the history of the numerical analysis of
Fredholm integral equations of the second kind 53
  Kendall Atkinson

Memoires on building a general purpose numerical algorithms library 73
  Brian Ford

Recent trends in high performance computing 93
  Jack J. Dongarra, Hans W. Meuer, Horst D. Simon, and Erich
  Strohmaier

Nonnegativity constraints in numerical analysis 109
  Donghui Chen and Robert J. Plemmons

On nonlinear optimization since 1959 141
  M. J. D. Powell

The history and development of numerical analysis in Scotland: a
personal perspective 161
  G. Alistair Watson

Remembering Philip Rabinowitz 179
  Philip J. Davis and Aviezri S. Fraenkel

My early experiences with scientific computation 187
  Philip J. Davis

Applications of Chebyshev polynomials: from theoretical kinematics to
practical computations 193
  Robert Piessens

Name Index 207

Subject Index 215

Some pioneers of extrapolation methods

Claude Brezinski

Laboratoire Paul Painlevé, UMR CNRS 8524, UFR de Mathématiques Pures et Appliquées, Université des Sciences et Technologies de Lille, 59655 - Villeneuve d'Ascq cedex, France.
Claude.Brezinski@univ-lille1.fr

Abstract. There are two extrapolation methods methods which are described in almost all numerical analysis books: Richardson's extrapolation method (which forms the basic ingredient for Romberg's method), and Aitken's Δ^2 process. In this paper, we consider the historical roots of these two procedures (in fact, the computation of π) with an emphasis on the pioneers of this domain of numerical analysis. Finally, we will discuss some more recent developments and applications.

Richardson's extrapolation method and Aitken's Δ^2 process are certainly the most well known methods for the acceleration of a slowly converging sequence. Both are based on the idea of extrapolation, and they have their historical roots in the computation of π.

We will first explain what extrapolation methods are, and how they lead to sequence transformations for accelerating the convergence. Then, we will present the history of Richardson's extrapolation method, of Romberg's method, and of Aitken's Δ^2 process, with an emphasis on the lives and the works of the pioneers of these topics.

The study of extrapolation methods and convergence acceleration algorithms now forms an important domain of numerical analysis having many applications; see [15, 24, 71, 77, 78, 80]. More details about its mathematical developments and its history could be found in [11, 13, 26, 34].

1 Interpolation, extrapolation, sequence transformations

Assume that the values of a function f are known at k distinct points x_i, that is

$$y_i = f(x_i), \quad i = 0, \ldots, k - 1.$$

Choose a function F_k depending on k parameters $a_0, \ldots, a_{k-1}$, and belonging to some class of functions $\mathcal{F}_k$ (for example polynomials of degree $k - 1$).

What is interpolation? Compute $a_0^e, \ldots, a_{k-1}^e$ solution of the system of equations (the meaning of the superscript $.^e$ will appear below)

$$F_k(a_0^e, \ldots, a_{k-1}^e, x_i) = y_i, \quad i = 0, \ldots, k - 1. \tag{1.1}$$

1

Then, for any $\forall x \in I = [\min_i x_i, \max_i x_i]$, we say that f has been *interpolated* by $F_k \in \mathcal{F}_k$, and we have $F_k(a_0^e, \ldots, a_{k-1}^e, x) \simeq f(x)$. Moreover, if $f \in \mathcal{F}_k$, then, for all x, $F_k(a_0^e, \ldots, a_{k-1}^e, x) = f(x)$.

What is extrapolation? Now choose $x^e \notin I$, and compute

$$y^e = F_k(a_0^e, \ldots, a_{k-1}^e, x^e),$$

where the coefficients a_i^e are those computed as the solution of the system (1.1). The function f has been *extrapolated* by $F_k \in \mathcal{F}_k$ at the point x^e, and $y^e \simeq f(x^e)$. Again, if $f \in \mathcal{F}_k$, then $F_k(a_0^e, \ldots, a_{k-1}^e, x^e) = f(x^e)$.

What is a sequence transformation? Assume now, without restricting the generality, that we have an infinite decreasing sequence of points $x_0 > x_1 > x_2 > \cdots > x^*$ such that $\lim_{n \to \infty} x_n = x^*$. We set $y_i = f(x_i)$, $i = 0, 1, \ldots$, and we also assume that $\lim_{n \to \infty} y_n = \lim_{n \to \infty} f(x_n) = y^*$. For any fixed n, compute $a_0^{(n)}, \ldots, a_{k-1}^{(n)}$ solution of the system

$$F_k(a_0^{(n)}, \ldots, a_{k-1}^{(n)}, x_{n+i}) = y_{n+i}, \quad i = 0, \ldots, k-1.$$

Then, compute $F_k(a_0^{(n)}, \ldots, a_{k-1}^{(n)}, x^*)$. This value obviously depends on n and, for that reason, it will be denoted by z_n. Then, the sequence (y_n) has been transformed into the new sequence (z_n), and $T : (y_n) \longmapsto (z_n)$ is called a *sequence transformation*. As we can see, it's a kind of *moving* extrapolation or, if one prefers, it is a sequence of extrapolated values based on different points. Obviously, if $f \in \mathcal{F}_k$, then, for all n, $z_n = y^*$. Instead of a decreasing sequence (x_n), we can take an increasing one, for example $(x_n = n)$.

A sequence transformation can be defined without any reference to a function f, but only to a sequence (y_n). An important concept is the notion of *kernel* of a sequence transformation: it is the set $\mathcal{K}_T$ of sequences (y_n) such that, for all n, $z_n = y^*$. If (y_n) converges, y^* is its limit, otherwise it is called its *antilimit*.

For readers who are not familiar with the topics of interpolation, extrapolation, and sequence transformations, these notions will be explained again at the beginning of Section 3 via the construction of Aitken's Δ^2 process.

When are extrapolation and sequence transformations powerful? As explained above, extrapolation and sequence transformations are based on the choice of the the class of functions $\mathcal{F}_k$. Thus, if the function f to be extrapolated behaves like a function of $\mathcal{F}_k$, the extrapolated value y^e will be a good approximation of $f(x^e)$. Similarly, the sequence (z_n) will converge to y^* faster than the sequence (y_n), that is $\lim_{n \to \infty} (z_n - y^*)/(y_n - y^*) = 0$, if (y_n) is close, in a sense to be defined (an open problem), to $\mathcal{K}_T$.

For each sequence transformation, there are sufficient conditions so that (z_n) converges to the same limit as (y_n), but faster. It was proved that a universal sequence transformation able to accelerate the convergence of *all* converging sequences cannot exist [25]. This is even true for some classes of sequences such as the monotonically decreasing ones. This negative result does not mean that a particular sequence belonging to such a class cannot be accelerated, but that an algorithm for accelerating all of them cannot exist.

2 Richardson's extrapolation

The procedure named after Richardson consists in extrapolation at $x^e = 0$ by a polynomial. It is carried out by means of the Neville-Aitken scheme for the recursive computation of interpolation polynomials.

2.1 First contributions

The number π can be approximated by considering polygons with n sides inscribed into and circumscribed to a circle. With $n = 96$, Archimedes obtained two significant figures. He also proved, by geometrical arguments, that the area of a circle is equal to $rp/2$, where r is its radius and p its perimeter.

In 1596, Adriaan van Roomen, also called Adrianus Romanus (Leuven, 1561 - Mainz, 1615), obtained 15 figures with $n = 2^{30}$ while, in 1610, Ludolph van Ceulen (Hildesheim, 1540 - Leiden, 1610), with the help of his student Pieter Cornelisz (Amsterdam, 1581 - The Hague, 1647), gave 36 figures by using $n = 2^{62}$. According to the Dutch mathematician David Bierens de Haan (1822 - 1895), these values were, in fact, obtained at about the same time. Van Ceulen's result was carved on his tombstone in the St. Peters church in Leiden. His work was continued in 1616 by Philips van Lansbergen (Ghent, 1561 - Middleburg, 1632) who held him in high esteem. He was a minister who published books on mathematics and astronomy where he supported Copernic's theories. However, he did not accept Kepler's theory of elliptic orbits. He suggested the approximation

$$2\pi \simeq p_{2n} + (p_{2n} - p_n)\frac{p_{2n} - 4}{p_n},$$

where p_n is the perimeter of the regular n-gons inscribed in the unit circle. He obtained π with 28 exacts decimal figures. The Dutch astronomer Willebrord Snel van Royen (Leiden, 1580 - Leiden, 1626), known as Snellius, was the introducer of the method of triangulation for measuring the length of the meridian. He also proposed, in 1621, the lower and upper bounds

$$\frac{3p_{2n}^2}{2p_{2n} + p_n} < 2\pi < \frac{p_{2n}(p_{2n} + 2p_n)}{3p_n}.$$

These formulae were preparing the ground for the next step.

2.2 C. Huygens

In 1654, Christiaan Huygens (The Hague, 1629 - id., 1695), in his *De Circuli Magnitudine Inventa* [33], proved 16 theorems or lemmas on the geometry of inscribed and circumscribed regular polygons. In particular, he gave the difference between the areas of the polygon with $2n$ sides and that with n sides. Let a_c and a_n be the areas of the circle and of the n-gons, respectively. He proved that (see [50] for an analysis of the proof)

$$
\begin{aligned}
a_c &= a_n + (a_{2n} - a_n) + (a_{4n} - a_{2n}) + (a_{8n} - a_{4n}) + \cdots \\
&> a_n + (a_{2n} - a_n) + (a_{2n} - a_n)/4 + (a_{2n} - a_n)/16 + \cdots \\
&= a_{2n} + (a_{2n} - a_n)(1/4 + 1/16 + \cdots).
\end{aligned}
$$

Then Huygens refers to Archimedes when stating that the sum of this geometric series is $1/3$, thus leading to

$$
a_c > a_{2n} + \frac{a_{2n} - a_n}{3}.
$$

We have

$$
a_n = \frac{1}{2} n \sin\left(\frac{2\pi}{n}\right) < \pi < A_n = n \tan\left(\frac{\pi}{n}\right),
$$

where A_n is the area of the circumscribed n-gons. Series expansions of trigonometric functions were not available to Huygens. However

$$
a_n = \pi - \frac{2\pi^3}{3n^2} + \frac{2\pi^5}{15n^4} - \frac{4\pi^7}{315n^6} + \cdots,
$$

and so Huygens' lower bound is such that

$$
a_{2n} + \frac{a_{2n} - a_n}{3} = \pi - \frac{8\pi^5}{15 \cdot 16n^4} + \frac{16\pi^7}{63 \cdot 64n^6} - \cdots
$$

Similarly

$$
A_n = \pi + \frac{\pi^3}{2n^2} + \frac{2\pi^5}{15n^4} + \frac{17\pi^7}{315n^6} + \cdots
$$

and it follows

$$
A_{2n} + \frac{A_{2n} - A_n}{3} = \pi - \frac{8\pi^5}{15 \cdot 16n^4} - \frac{68\pi^7}{63 \cdot 64n^6} - \cdots,
$$

which is also a lower bound for π, but slightly poorer than the previous one.

Thus, Huygens' formulae for lower bounds are exactly those obtained by the first step of Richardson's extrapolation method. Moreover, in order to obtain an upper bound, he proposed

$$
A_n - \frac{A_n - a_n}{3} = \pi + \frac{2\pi^5}{15n^4} + \cdots,
$$

whose error is bigger than for the lower bounds but uses only polygons with n sides instead of $2n$. With $n = 2^{30}$, this last formula doubles (up to 35) the number of exact digits of π. In letters to Frans van Schooten (Leiden, 1615 - id., 1660) and Daniel Lipstorp (1631 - 1684), he claimed that he was able to triple the number of exact decimals. Therefore, Huygens achieved approximations of π which are even better than those given by Richardson's extrapolation!

Huygens' method was later used by Jacques Frédéric Saigey in 1856 and 1859 [61]. He considered the three approximations

$$A'_n = A_{2n} + \frac{1}{3}(A_{2n} - A_n)$$

$$A''_n = A'_{2n} + \frac{1}{15}(A'_{2n} - A'_n)$$

$$A'''_n = A''_{2n} + \frac{1}{63}(A''_{2n} - A''_n)$$

which are similar to those that will be given later by Romberg in the context of accelerating the convergence of the trapezoidal rule.

Saigey was born in Montbéliard in 1797. He studied at the École Normale Supérieure in Paris, but the school was closed in June 1822 by the regime of Louis XVIII. Saigey became the secretary of Victor Cousin and helped him to publish the volume V of Descartes' complete works. Then, he became one of the main editors of the journal *Bulletin des Sciences Mathématiques*. He published several papers in mathematics and physics, but he was mostly known for his elementary treatises and memoranda which had several editions. He died in Paris in 1871.

In 1903, Robert Moir Milne (1873 - ?) applied Huygens' ideas for computing π [43], as also did Karl Kommerell (1871 - 1948) in his book of 1936 [36]. As explained in [76], Kommerell can be considered as the real discoverer of Romberg's method since he suggested the repeated use of Richardson's rule, although it was in a different context.

2.3   L.F. Richardson

In 1910, Lewis Fry Richardson (1881 - 1953) suggested to eliminate the first error term in the central differences formulæ given by William Fleetwood Sheppard (Sydney, 1863 - 1936) [69] by using several values of the stepsize. He wrote [52]

... the errors of the integral and of any differential expressions derived from it, due to using the simple central differences of §1.1 instead of the differential coefficients, are of the form

$$h^2 f_2(x, y, z) + h^4 f_4(x, y, z) + h^6 f_6(x, y, z) + \&tc.$$

Consequently, if the equation be integrated for several different values of h, extrapolation on the supposition that the error is of this form will give numbers very close to the infinitesimal integral.

In 1927, Richardson called this procedure the *deferred approach to the limit* [55]. Let us quote him

Confining attention to problems involving a single independent variable
x, let h be the "step", that is to say, the difference of x which is used in
the arithmetic, and let $\phi(x, h)$ be the solution of the problem in differ-
ences. Let $f(x)$ be the solution of the analogous problem in the infinites-
imal calculus. It is $f(x)$ which we want to know, and $\phi(x, h)$ which is
known for several values of h. A theory, published in 1910, but too brief
and vague, has suggested that, if the differences are "centered" then

$$\phi(x, h) = f(x) + h^2 f_2(x) + h^4 f_4(x) + h^6 f_6(x)...\text{to infinity...} \quad (1)$$

odd powers of h being absent. The functions $f_2(x), f_4(x), f_6(x)$ are usu-
ally unknown. Numerous arithmetical examples have confirmed the ab-
sence of odd powers, and have shown that it is often easy to perform the
arithmetic with several values of h so small that $f(x) + h^2 f_2(x)$ is a good
approximation to the sum to infinity of the series in (1).
If generally true, this would be very useful, for it would mean that if we
have found two solutions for unequal steps h_1, h_2, then by eliminating
$f_2(x)$ we would obtain the desired $f(x)$ in the form

$$f(x) = \frac{h_2^2 \phi(x, h_1) - h_1^2 \phi(x, h_2)}{h_2^2 - h_1^2}. \quad (2)$$

This process represented by the formula (2) will be named the "h^2-extra-
polation".
If the difference problem has been solved for three unequal values of h it is
possible to write three equations of the type (1) for h_1, h_2, h_3, retaining
the term $h^4 f_4(x)$. Then $f(x)$ is found by eliminating both $f_2(x)$ and
$f_4(x)$. This process will be named the "h^4-extrapolation".

Let us mention that Richardson referred to a paper by Nikolai Nikolaevich
Bogolyubov (Nijni-Novgorod, 1909 - Moscow, 1992) and Nikolai Mitrofanovich
Krylov (Saint Petersburg, 1879 - Moscow, 1955) of 1926 where the deferred
approach to the limit can already be found [7].

In the same paper, Richardson used this technique for solving a 6th order
differential eigenvalue problem. Richardson extrapolation consists in fact in computing the value at 0, denoted by $T_k^{(n)}$, of the interpolation polynomial of the degree at most k which passes through the points $(x_n, S(x_n)), \ldots, (x_{n+k}, S(x_{n+k}))$.
Thus, using the Neville-Aitken scheme for these interpolation polynomials, the
numbers $T_k^{(n)}$ can be recursively computed by the formula

$$T_{k+1}^{(n)} = \frac{x_{n+k+1} T_k^{(n)} - x_n T_k^{(n+1)}}{x_{n+k+1} - x_n}, \qquad k, n = 0, 1, \ldots \quad (2.1)$$

with $T_0^{(n)} = S(x_n)$ for $n = 0, 1, \ldots$.

Extensions of the Richardson extrapolation process are reviewed in [15, 71],
and many applications are discussed in [39].

Lewis Fry (the maiden name of his mother) Richardson was born on October 11, 1881 in Newcastle upon Tyne, England, the youngest of seven children in a Quaker family. He early showed an independent mind and had an empirical approach. In 1898, he entered the Durham College of Science where he took courses in mathematics, physics, chemistry, botany, and zoology. Then, in 1900, he went to King's College in Cambridge, and followed the physics lectures of Joseph John Thompson (Cheetham Hill near Manchester, 1856 - Cambridge, 1940), the discoverer of the electron. He graduated with a first-class degree in 1903. He spent the next ten years holding a series of positions in various academic and industrial laboratories. When serving as a chemist at the National Peat Industry Ltd., he had to study the percolation of water. The process was described by the Laplace equation on an irregular domain and Richardson used finite differences, and extrapolation. But it was only after much deliberation and correspondence that his paper was accepted for publication [52]. He submitted this work for a D.Sc. and a fellowship at Cambridge, but it was rejected. The ideas were too new, and the mathematics were considered as "approximate mathematics"! Hence, Richardson never worked in any of the main academic research centers. This isolation probably affected him. For some time, he worked with the well-known statistician Karl Pearson (London, 1857 - Coldharbour, 1936), and became to be interested in "living things".

In 1913, Richardson became Superintendent of the Eskdalemuir Observatory in southern Scotland. He had no experience in meteorology, but was appointed to bring some theory in its understanding. He again made use of finite differences. Although he was certainly aware of the difficulty of the problem since he estimated at 64.000 the number of people that have to be involved in the computations in order to obtain the prediction of tomorrow's weather before day actually began, it seems that he did not realize that the problem was ill-conditioned. He also began to write his book on this topic [53]. The quote at the end of its preface is amusing.

This investigation grew out of a study of finite differences and fist took shape in 1911 as the fantasy which is now relegated to Chap. 11/2. Serious attention to the problem was begun in 1913 at Eskdalemuir Observatory with the permission and encouragement of Sir Napier Shaw, then Director of the Meteorological Office, to whom I am greatly indebted for facilities, informations and ideas. I wish to thank Mr. W.H. Dines, F.R.S., for his interest in some early arithmetical experiments, and Dr. Crichton Mitchell, F.R.S.E., for some criticisms of the first draft. The arithmetical reduction of the balloon, and other observations, was done with much help from my wife. In May 1916 the manuscript was communicated by Sir Napier Shaw to the Royal Society, which generously voted £100 towards to cost of its publication. The manuscript was revised and the detailed example of Chap. IX was worked out in France in the intervals of transporting wounded in 1916-1918. During the battle of Champagne in April 1917 the working copy was sent to the rear, where it became lost, to be re-discovered some months later under a heap of coal. In 1919, as printing was delayed by the legacy of the war, various excrescences were removed for separate publication, and an introductory example was added. This was done at Benson, where I had again the good fortune to be able to discuss the hypotheses with Mr. W.H. Dines. The whole work has been thoroughly revised in 1920, 1921. As the cost of printing had by this time much increased, an application was made to Dr. G.C. Simpson, F.R.S., for a further grant in aid, and the sum of fifty pounds was provided by the Meteorological Office.

As Richardson wrote, on May 16, 1916 he resigned and joined the Friends' Ambulance Unit (a Quaker organisation) in France. He began to think about the causes of wars and how to prevent them. He suggested that the animosity between two countries could be measured, and that some differential equations are involved into the process. He published a book on these ideas [54], and then returned to weather prediction.

Along the years, Richardson made important contributions to fluid dynamics, in particular to eddy-diffusion in the atmosphere. The so-called "Richardson number" is a fundamental quantity involving gradients of temperature and wind velocity.

In 1920, he became a Lecturer in mathematics and physics at the Westminster Training College, an institution training prospective school teachers up to a bachelor's degree. In 1926, he changed again his field of research to psychology where he wanted to apply the ideas and the methods of mathematics and physics. He established that many sensations are quantifiable, he found methods for measuring them, and modelled them by equations. The same year, he was elected as a Fellow of the Royal Society of London.

Richardson left the Westminster Training College in 1929 for the position of Principal at the Technical College in Paisley, an industrial city near Glasgow. Although he had to teach sixteen hours a week, he continued his research but came back to the study of the causes of wars and their prevention. He prepared a model for the tendencies of nations to prepare for wars, and worked out its applications using historical data from the previous conflicts. He also made predictions for 1935, and showed that the situation was unstable, which could only be prevented by a change in the nation's policies. Richardson wanted to "see whether there is any statistical connection between war, riot and murder". He began to accumulate such data [57], and decided to search for a relation between the probability of two countries going to war and the length of their common border. To his surprise, the lengths of the borders were varying from one source to another. Therefore, he investigated how to measure the length of a border, and he realized that it highly depends on the length of the ruler. Using a small ruler allows to follow more wiggles, more irregularities, than a long one which cuts the details. Thus, the smaller the ruler, the larger the result. The relation between the length of the border and that of the ruler leads to a new mathematical measure of wiggliness. At that time, Richardson's results were ignored by the scientific community, and they were only published posthumously [58]. Today, they are considered to be at the origin of fractals.

In 1943, Richardson and his wife moved to their last home at Kilmun, 25 miles from Glasgow. He returned to his research on differential equations, and solved the associated system of linear equations by the so-called Richardson's method [56]. He mentioned that the idea was suggested to him in 1948 by Arnold Lubin. At home, Richardson was also constructing an analogous computer for his meteorological computations. He died on September 30, 1953 in Kilmun.

Richardson was a very original character whose contributions to many different fields were prominent but, unfortunately, not appreciated at their real values at his epoch; see [32] for details.

2.4 W. Romberg

Let us now come to the procedures for improving the accuracy of the trapezoidal rule for computing approximations to a definite integral. If the function to be integrated is sufficiently differentiable, the error of the trapezoidal rule is given by the Euler-Maclaurin expansion. In 1742, Colin Maclaurin (Kilmodan, 1698 - Edinburgh, 1746) [38] showed that the precision could be improved by linear combinations of the results obtained with various stepsizes. His procedure can be interpreted as a preliminary version of Romberg's method; see [21] for a discussion.

In 1900, Sheppard used an elimination strategy in the Euler-Maclaurin quadrature formula, with $h_n = r_n h$ and $1 = r_0 < r_1 < r_2 < \cdots$, for producing a better approximation [70]. In 1952, Mario Salvadori (Rome, 1907 - 1997), an architect and structural engineer, and Melvin L. Baron (1927 - 1997), a civil engineer, proposed to use Richardson's deferred approach to the limit for improving the trapezoidal rule [62]. This new approximation was obtained as a linear combination of the initial results.

In 1955, Werner Romberg was the first to use repeatedly an elimination approach for improving the accuracy of the trapezoidal rule [59]. He gave the well known formula

$$T_{k+1}^{(n)} = \frac{4^{k+1} T_k^{(n+1)} - T_k^{(n)}}{4^{k+1} - 1},$$

where $T_0^{(n)}$ is the result obtained by the trapezoidal rule with the stepsize $h_0/2^n$. In his paper, Romberg refers to the book of Lothar Collatz (Arnsberg, Westfalia, 1910 - Varna, 1990) of 1951 [22].

In 1960, Eduard L. Stiefel (1909 - 1978), in his inaugural address as the President of the IFIP congress in Munich, draws a line from Archimedes to Romberg. The procedure became widely known after the rigorous error analysis given in 1961 by Friedrich L. Bauer (born 1924 in Regensburg) [6] and the synthesis of Stiefel [75]. Romberg's derivation of his process was mainly heuristic. It was proved by Pierre-Jean Laurent in 1963 [37] that the process comes out, in fact, from the Richardson process when taking $x_n = h_n^2$ and $h_n = h_0/2^n$. Laurent also gave the condition on the sequence (h_n) that there exists $\alpha < 1$ such that $\forall n, h_{n+1}/h_n \leq \alpha$ in order that the sequences $(T_k^{(n)})$ tend to the exact value of the definite integral to be computed either when k or n tends to infinity. The case of a harmonic sequence of steps is studied in [23, p. 52]. Romberg's work on the extrapolation of the trapezoidal rule has been continued Tore Håvie for less regular integrands [28].

Werner Romberg was born on May 16, 1909 in Berlin. In 1928, he started to study physics and mathematics in Heidelberg where the Nobel laureate Philip Lenard (Pozsony, Pressburg, 1862 - Messelhausen, 1947) was still quite influential. After two years, Romberg decided to go to the Ludwig-Maximilians University in Munich. He followed the mathematics courses of Constantin Carathéodory (Berlin, 1873 - Munich, 1950) and Oskar Perron (Frankenthal, Pfalz, 1880 - Munich, 1975), and had physics lectures by Arnold Sommerfeld (Königsberg, 1868 - Munich, 1951), who became his advisor. In 1933, he defended his thesis *Zur Polarisation des Kanalstrahllichtes* (On the polarization of canal jet rays). The same year, he had to leave Germany and went to the USSR. He stayed at the Department of Physics and Technology in Dnepropetrovsk from 1934 to 1937 as a theoretical physicist. He was briefly at the Institute of Astrophysics in Prag in 1938, but he had to escape from there. Then, he got a position in Oslo in the autumn of 1938 as the assistant of the physicist Egil Andersen Hylleraas (Engerdal, 1898 - 1965). He also worked for a short period with Johan Holtsmark (1894 - 1975), who built a Van de Graaff generator (the second one in Europe and the first particle accelerator in Scandinavia) for nuclear disintegration between 1933 and 1937 at Norwegian Institute of Technology (NTH) in Trondheim. Romberg had again to escape for some time to Uppsala during the German occupation of Norway. In 1949, he joined the NTH in Trondheim as an associate professor in physics. In 1960, he was appointed head of the Applied Mathematics Department at the NTH. He organized a teaching program in applied mathematics, and began to build a research group in numerical analysis. He was strongly involved in the introduction of digital computers in Norway, and in the installation of the first computer (GIER) at NTH. He became a Norwegian citizen and stayed Norwegian until the end of his life.

In 1968, Romberg came back to Heidelberg where he accepted a professorship. He built up a group in numerical mathematics, at that time quite underdeveloped in Heidelberg, and was the head of the Computing Center of the University from 1969 to 1975. Romberg retired in 1978, and died on February 5, 2003.

3   Aitken's process and Steffensen's method

Let (S_n) be a sequence of scalars converging to S. The most popular nonlinear acceleration method is certainly Aitken's Δ^2 process which consists in building a new sequence (T_n) by

$$T_n = \frac{S_n S_{n+2} - S_{n+1}^2}{S_{n+2} - 2S_{n+1} + S_n}, \quad n = 0, 1, \ldots \tag{3.1}$$

For deriving this formula, Aitken assumed that he had a sequence (S_n) of the form

$$S_n = S + \alpha \lambda^n, \quad n = 0, 1, \ldots \tag{3.2}$$

with $\lambda \neq 1$, and he wanted to compute S (the limit of the sequence if $|\lambda| < 1$, its antilimit otherwise). Then, $\Delta S_n = \alpha \lambda^n (\lambda - 1)$, and $\lambda = \Delta S_{n+1}/\Delta S_n$. It follows

$$S = S_n - \frac{\Delta S_n}{(1 - \lambda)} = S_n - \frac{\Delta S_n}{(1 - \Delta S_{n+1}/\Delta S_n)} = \frac{S_n S_{n+2} - S_{n+1}^2}{S_{n+2} - 2S_{n+1} + S_n}.$$

If (S_n) has not the form (3.2), the preceding formula can still be used, but the result is no longer equal to S. It depends on n, and it is denoted by T_n as in (3.1). This construction of Aitken's process illustrates how interpolation, extrapolation, and sequence transformations are related. Indeed, let (S_n) be any sequence. We are looking for S, α and λ satisfying the interpolation conditions $S_i = S + \alpha \lambda^i$ for $i = n, n + 1, n + 2$. Then, the unknown S is taken as the limit when n tends to infinity of the model sequence $(S + \alpha \lambda^n)$. This is an extrapolation process. But, since the value of S obtained in this procedure depends of n, it has been denoted by (S_n), and, thus, the given sequence (S_n) has been transformed into the new sequence (T_n).

Thus, by construction, the kernel of Aitken's process consists in sequences of the form (3.2), or, in other terms, of sequences satisfying a first order linear difference equation

$$a_0(S_n - S) + a_1(S_{n+1} - S) = 0, \quad n = 0, 1, \ldots$$

with $a_0 + a_1 \neq 0$.

If (S_n) is linearly converging, i.e. if a number $\lambda \neq 1$ exists such that

$$\lim_{n \to \infty} \frac{S_{n+1} - S}{S_n - S} = \lambda,$$

then (T_n) converges to S faster than (S_n). This result illustrates the fact mentioned above that sequences *not too far away from the kernel* (in a meaning to be defined) are accelerated. Acceleration is also obtained for some subclasses of sequences satisfying the preceding property with $\lambda = 1$ (logarithmically converging sequences).

In a paper of 1937 [2], Aitken used his process for accelerating the convergence of the power method (Rayleigh quotients) for computing the dominant eigenvalue of a matrix. A section is entitled *The δ^2-process for accelerating convergence*, and, on pages 291–292, he wrote

For practical computation it may be remembered by the following memoria technica: product of outers minors *[minus]* square of middle, divided by sum of outers minus double of middle.

Aitken's paper [2] also contains almost all the ideas that will be developed later by Heinz Rutishauser (Weinfelden, 1918 - 1970) in his QD-algorithm [60]. Notice that Formula (3.1) is numerically unstable, and that one should prefer the following one

$$T_n = S_{n+1} + \frac{(S_{n+1} - S_n)(S_{n+2} - S_{n+1})}{(S_{n+1} - S_n) - (S_{n+2} - S_{n+1})}. \tag{3.3}$$

It is well known that the fixed point iterative method due to Johan Frederik Steffensen (1873-1961) in 1933 is based on Aitken's process. However, Steffensen does not quote Aitken in his paper, and his discovery seems to have been obtained independently. Consider the computation of x such that $x = f(x)$ and the iterations $x_{\nu+1} = f(x_\nu)$. Steffensen writes [74]

In the linear interpolation formula with divided differences

$$f(x) = f(a_0) + (x - a_0)f(a_0, a_1) + (x - a_0)(x - a_1)f(x, a_0, a_1) \quad (5)$$

we put $a_\nu = x_\nu$ and obtain

$$f(x) = x_1 + (x - x_0)\frac{x_1 - x_2}{x_0 - x_1} + R_1$$

where

$$R_1 = (x - x_0)(x - x_1)f(x, x_0, x_1). \quad (6)$$

Replacing, on the left of (6), $f(x)$ by x, we have

$$x = x_1 + (x - x_0)\frac{x_1 - x_2}{x_0 - x_1} + R_1$$

and solving for x, as if R_1 were a constant, we obtain after a simple reduction

$$x = x_0 - \frac{(\Delta x_0)^2}{\Delta^2 x_0} + R \quad (7)$$

where

$$R = -(x - x_0)(x - x_1)\frac{\Delta x_0}{\Delta^2 x_0}f(x, x_0, x_1). \quad (8)$$

If $f(x)$ possesses a continuous second derivative, the remainder may be written

$$R = -\frac{1}{2}(x - x_0)(x - x_1)\frac{\Delta x_0}{\Delta^2 x_0}f''(\xi). \quad (9)$$

The formula (7) is the desired result. The approximation obtained may often be estimated by (9), but we shall make no use of this formula, preferring to test the result by other methods. We shall therefore use as working formula the approximation

$$x = x_\nu - \frac{(\Delta x_\nu)^2}{\Delta^2 x_\nu} \tag{10}$$

where, according to the remark made above, x_ν may be any element of the sequence.

Then, Steffensen gave several numerical examples where, after 3 iterations, he restarted them from the approximation given by (10). In a footnote to page 64 (the first page of his paper), he wrote

The present notes had already been written when a paper by H. Holme appeared in this journal (1932, pp. 225-250), covering to some extend the same ground. Mr. Holme's treatment of the subject differs, however, so much from mine that I think there is room for both.

In his paper [31], Harald Holme was solving a fixed point problem due to Birger Øivind Meidell (1882-1958) and related to the interest rate of loans [41]. He used linear interpolation passing through 3 consecutive iterates, and he obtained a method quite close to Steffensen's but different from it.

3.1   Seki Takakazu

In the fourth volume of his book *Katsuyō Sanpō*, published in 1674, Seki Takakazu considered the perimeters c_i of the polygons with 2^i sides inscribed into a circle of diameter 1. For deriving a better approximation of π, he used a method called *Yenri*, which means *principle* (or *theory*) *of the circle*, and consists in the formula

$$c_{16} + \frac{(c_{16} - c_{15})(c_{17} - c_{16})}{(c_{16} - c_{15}) - (c_{17} - c_{16})}.$$

This is exactly Aitken's Δ^2 process (as given by (3.3)) which leads to 12 exact decimal digits while c_{17} has only 10. With

$c_{15} = \underline{3.1415926}487769856708$

$c_{16} = \underline{3.1415926}523565913571$

$c_{17} = \underline{3.1415926}532889027755,$

Seki obtained $\underline{3.14159265359}$ ($\pi = 3,14159265358979323846\ldots$). His result is, in fact, exact to 16 places. Seki did not explain how he got his formula but, probably, setting $a = c_{15}, b = c_{16} = a + ar$, and $c = c_{17} = a + ar + ar^2$, he obtained [30]

$$b + \frac{(b-a)(c-b)}{(b-a) - (c-b)} = \frac{a}{1-r} = a + ar + ar^2 + ar^3 + \cdots$$

The same method was used by his student Takaaki Takebe (1661 - 1716), who developed it further. Seki also studied how to compute an arc of a circle, given the chord, and he used again his formula for improving his first approximations.

Seki Takakazu is considered as the greatest Japanese mathematician. He was born in Fujioka in 1637 or in 1642. He was later adopted by the Seki family. However, little is known about his life, but it seems that he was self-educated and an infant prodigy in mathematics. In his book mentioned above, he introduced a notation for representing unknowns and variables in equations, and he solved fifteen problems which had been posed three years earlier by Kazuyuki Sawaguchi (it was the habit to end a book by open problems). He anticipated many discoveries of western mathematicians: determinants (1683, ten years before Leibniz) for solving systems of 2 or 3 linear equations, Bernoulli numbers, Newton-Raphson method, and Newton interpolation formula. He studied the solution of equations with negative and positive zeros, and, in 1685, he solved the cubic equation $30 + 14x - 5x^2 - x^3 = 0$ by the same method as Horner a hundred years later. He was also interested in magic squares, and Diophantine equations. He died in 1708.

3.2 A.C. Aitken

The sequence transformation defined by (3.1) was stated by Alexander Craig Aitken in 1926 [1] who used it for accelerating the convergence of Daniel Bernoulli's method of 1728 for the computation of the dominant zero z_1 of the polynomial $a_0 z^n + \cdots + a_{n-1}z + a_n$. The method imagined by Bernoulli consists in considering the sequence $Z_1(t) = f(t+1)/f(t)$ generated from the recursion $a_0 f(t+n) + \cdots + a_n f(t) = 0$, and whose limit is z_1 (assuming that all other zeros of the polynomial have a modulus strictly smaller than $|z_1|$). With this condition, Aitken writes

$\Delta Z_1(t)$ tends to become a geometric sequence... of common ratio z_2/z_1. Hence the derivations of $Z_1(t)$ from z_1 will also tend to become a geometric sequence with the same common ratio. Thus a further approximate solution is suggested, viz.

$$\frac{z_1 - Z_1(t+2)}{z_1 - Z_1(t+1)} = \frac{\Delta Z_1(t+1)}{\Delta Z_1(t)}$$

and solving for z_1 we are led to investigate the derived sequence

$$Z_1^{(1)}(t) = \frac{\begin{vmatrix} Z_1(t+1) & Z_1(t+2) \\ Z_1(t) & Z_1(t+1) \end{vmatrix}}{\Delta^2 Z(t)}. \tag{8.2}$$

This is exactly (3.1). Aitken claims that this new sequence converges geometrically with the ratio $(z_2/z_1)^2$ or z_3/z_1, and that the process can be repeated on the sequence $(Z_1^{(1)}(t)$. In a footnote, he says that *Naegelsbach, in the course of a very detailed investigation of Fürstenau method of solving equations, obtains the formulæ (8.2) and (8.4), but only incidentally*. The reference for the work of Eduard Fürstenau is [27]. It must be pointed out that, on page 22 of his second paper [44], Hans von Naegelsbach (1838 - ?) gave the stable formulation (3.3) of the process.

The process was also given by James Clerk Maxwell (Edinburgh, 1831 - Cambridge, 1879) in his *Treatise on Electricity and Magnetism* of 1873 [40]. However, neither Naegelsbach nor Maxwell used it for the purpose of acceleration. Maxwell wanted to find the equilibrium position of a pointer oscillating with an exponentially damped simple harmonic motion from three experimental measurements (as in (3.2)).

Aitken was born in Dunedin, New Zealand, on April 1st, 1895. He attended Otago's High School from 1908 to 1912, where he was not particularly brillant. But, at the age of 15, he realized that he had a real power in mental calculations, and that his memory was extraordinary. He was able to recite the first 1000 decimals of π, and to multiply two numbers of nine digits in a few seconds [72]. He also knew the Aeneid by heart. He was also very good at several sports and began to study violin. He studied mathematics, French and Latin at the University of Otago in 1913 and 1914. It seems that the professor of mathematics there, David J. Richards, a "temperamental, eccentric Welshman", was lacking of the power to communicate his knowledge to the students, and Aitken's interest in mathematics lowered. Richards was trained as an engineer as well as mathematician, and was working as an engineer in Newcastle prior to his appointment to the Chair of Mathematics at Otago in 1907, where he stayed until 1917.

Aitken volunteered in the Otago infantry during World War I, and he took part in the Gallipoli landing and in the campaign in Egypt. Then, he was commissioned in the north of France, and was wounded in the shoulder and foot during the battle on the river Somme. Did he met Richardson at this time? After a stay in a London hospital, he was invalided home in 1917, and spent one year of recovering in Dunedin where he wrote a first account of his memoirs published later [4].

Aitken resumed his studies at Otago University, and graduated with first class honours in languages, but only with second ones in mathematics. He married Winifred Betts in 1920, and became a school teacher at his old Otago High School. Richards' successor in the Chair of Mathematics, Robert John Tainsh Bell was born in 1877. He graduated from the University of Glasgow in 1898, and was appointed Lecturer there three years later. He was awarded a D.Sc. in 1911, and was appointed Professor of Pure and Applied Mathematics at Otago University in 1919. Bell was the only staff member in the Mathematics Department, lecturing five days a week, each day from 8.00 am to 1.00 pm. He retired in 1948, and died in 1963. When Bell required an assistant he called on Aitken. He encouraged him to apply for a scholarship for studying with Edmund Taylor Whittaker (Southport, 1873 - Edinburgh, 1956) at Edinburgh. Aitken left New Zealand in 1923. His Ph.D. on the smoothing of data, completed in 1925, was considered so outstanding that he was awarded a D.Sc. for it. The same year, Aitken was appointed as a Lecturer at the University of Edinburgh where he stayed for the rest of his life. But, the efforts for obtaining his degree led him to a first severe breakdown in 1927, and then he was periodically affected by such crisis. They were certainly in part due to his fantastic memory which did not fade with time, and he was always remembering the horrors he saw during the war [5] (see also the biographical introduction by Peter C. Fenton given in this volume).

In 1936, Aitken became a Reader in statistics, and he was elected a Fellow of the Royal Society. In 1946, he was appointed to Whittaker's Chair in Mathematics. In 1956, he received the prestigious Gunning Victoria Jubilee Prize of the Royal Society of Edinburgh. In 1964, he was elected to the Royal Society of Literature. Aitken died in Edinburgh on November 3, 1967.

3.3   J.F. Steffensen

Since the life of Steffensen is not so well-known, let us give some informations about it following [49]. Johan Frederik Steffensen was born in Copenhagen on February 28, 1873. His father was the Supreme Judge of the Danish Army, and he, himself, took a degree in law at the University of Copenhagen. After a short period in Fredericia in the eastern part of the Jutland peninsula in Denmark, he returned to Copenhagen and began a career in insurance. He was self-taught in mathematics and, in 1912, he got a Ph.D. for a study in number theory. After three years as the managing director of a mutual life assurance society, he turned to teach insurance mathematics at the University of Copenhagen, first as a Lecturer and, from 1923 to 1943, as a Professor. However, he was still continuing to be interested in the world of affairs, and was an active member, and even the Chairman, of several societies. He published around 100 research papers in various fields of mathematics, and his book of 1927 [73] can be considered as one of the first books in numerical analysis since its chapters cover interpolation in one and several variables, numerical derivation, solution of differential equations, and quadrature. Steffensen loved English literature, especially Shakespeare. He died on December 20, 1961. For a photography of Steffensen, see [47].

3.4   D. Shanks

The idea of generalizing Aitken's process is due to Daniel Shanks. He wanted to construct a sequence transformation with a kernel consisting of sequences satisfying, for all n,

$$a_0(S_n - S) + a_1(S_{n+1} - S) + \cdots + a_k(S_{n+k} - S) = 0, \tag{3.4}$$

with $a_0 + a_1 + \cdots + a_k \neq 0$. Let us mention that a particular case of an arbitrary value of k was already studied by Thomas H. O'Beirne in 1947 [48]. Writing the relation (3.4) for the indexes $n, n+1, \ldots, n+k$ leads to

$$\begin{vmatrix} S_n - S & S_{n+1} - S & \cdots & S_{n+k} - S \\ S_{n+1} - S & S_{n+2} - S & \cdots & S_{n+k+1} - S \\ \vdots & \vdots & & \vdots \\ S_{n+k} - S & S_{n+k+1} - S & \cdots & S_{n+2k} - S \end{vmatrix} = 0.$$

After elementary manipulations on the rows and columns of this determinant, Shanks obtained

$$S = H_{k+1}(S_n)/H_k(\Delta^2 S_n), \tag{3.5}$$

where $\Delta^2 S_n = S_{n+2} - 2S_{n+1} + S_n$, and where H_k denotes a Hankel determinant defined as

$$H_k(u_n) = \begin{vmatrix} u_n & u_{n+1} & \cdots & u_{n+k-1} \\ u_{n+1} & u_{n+2} & \cdots & u_{n+k} \\ \vdots & \vdots & & \vdots \\ u_{n+k-1} & u_{n+k} & \cdots & u_{n+2k-2} \end{vmatrix}.$$

If (S_n) does not satisfy the relation (3.4), the ratio of determinants in the right hand side of (3.5) could nevertheless be computed but, in this case, the result obtained depends on n, and it is denoted by $e_k(S_n)$. Thus, the sequence (S_n) has been transformed into the new sequence $(e_k(S_n))$ for a fixed value of k or, more generally, into a set of new sequences depending on k and n. This sequence transformation is known as *Shanks' transformation*. Let us also mention that the same ratio of determinants was obtained by R.J. Schmidt in 1941 [63] while studying a method for solving systems of linear equations.

Dan Shanks was born on January 17, 1917 in Chicago. In 1937, he received a B.Sc. in physics. From 1941 to 1957, he was employed by the Naval Ordnance Laboratory (NOL) located in White Oak, Maryland. There, in 1949, he published a Memorandum describing his transformation [65]. Without having done any graduate work, he wanted to present this work to the Department of Mathematics of the University of Maryland as a Ph.D. thesis. But, he had first to complete the degree requirements before his work could be examined as a thesis. Hence, it was only in 1954 that he obtained his Ph.D. which was published in the *Journal of Mathematical Physics* [66]. Dan considered this paper as one of his best two (the second one was his computation of π to 100.000 decimals published with John Wrench [67]). After the NOL, Shanks worked at the David Taylor Model Basin in Bethesda where I met him in December 1976. Then, in 1977, he joined the University of Maryland where he stayed until his death on September 6, 1996. Dan served as an editor of *Mathematics of Computation* from 1959 until his death. He was very influential in this position which also led him to turn to number theory, a domain where his book became a classic [68]. More details on Shanks life and works can be found in [79].

3.5   P. Wynn

The application of Shanks' transformation to a sequence (S_n) needs the computation of the ratios of Hankel determinants given by (3.5). The numerators and the denominators in this formula can be computed separately by the well-known recurrence relation for Hankel determinants (a by-product of Sylvester's determinantal identity). This was the way O'Beirne and Shanks were implementing the transformation. However, in 1956, Peter Wynn (born in 1932) found a recursive algorithm for that purpose, the ε-algorithm [81], whose rules are

$$\varepsilon_{k+1}^{(n)} = \varepsilon_{k-1}^{(n+1)} + \frac{1}{\varepsilon_k^{(n+1)} - \varepsilon_k^{(n)}}, \quad k, n = 0, 1, \ldots$$

with $\varepsilon_{-1}^{(n)} = 0$ and $\varepsilon_0^{(n)} = S_n$, for $n = 0, 1, \ldots$.
These quantities are related to Shanks' transformation by

$$\varepsilon_{2k}^{(n)} = e_k(S_n),$$

and the quantities with an odd lower index satisfy $\varepsilon_{2k+1}^{(n)} = 1/e_k(\Delta S_n)$. When $k = 1$, Aitken's process is recovered. The proof makes use of Schweins' and Sylvester's determinantal identities that could be found, for example, in Aitken's small monograph [3].

Later, Wynn became Bauer's assistant in Mainz, then he went to Amsterdam, participating in the birth of ALGOL, and then he held several researcher's positions in the United States, Canada, and Mexico. Wynn's ε-algorithm is certainly the most important and well-known nonlinear acceleration procedure used so far. Wynn dedicated many papers to the properties and the applications of his ε-algorithm. With a vector generalization of it [82], he also opened the way to special techniques for accelerating the convergence of sequences of vectors. The ε-algorithm also provides a derivative free extension of Steffensen's method for the solution of systems of nonlinear equations [8] (see also [15]).

Let us mention the important connection between the ε-algorithm and Padé approximants (and, thus, also with continued fractions). Let f be a formal power series

$$f(x) = \sum_{i=0}^{\infty} c_i x_i.$$

If the ε-algorithm is applied to its partial sums, that is $S_n = \varepsilon_0^{(n)} = \sum_{i=0}^{n} c_i x_i$, then $\varepsilon_{2k}^{(n)} = [n + k/k]_f(x)$, the Padé approximant of f with a numerator of degree $n + k$ and a denominator of degree k, a property exhibited by Shanks [66]. This connection allowed Wynn to obtain a new relation, known as the *cross rule*, between 5 adjacent approximants in the Padé table [83]. However, the ε-algorithm and the cross rule give the values of the Padé approximants only at the point x where the partial sums S_n were computed, while knowing the coefficients of the numerators and the denominators of the Padé approximants allows to compute them at any point.

4   And now?

In the last twenty years, Richardson's and Romberg's methods, Aitken's process and the ε-algorithm have been extended to more general kernels, or to accelerate new classes of sequences. Very general extrapolation algorithms have been obtained; see, for example, [15, 71]. In particular, the E-algorithm, whose rules obviously extend those of the Richardson process, was devised almost simultaneously by different people in different contexts [9, 29, 42, 64]. These procedures are now used in many physical applications [78, 20]. An important new field of investigation is the connection between some convergence acceleration algorithms and integrable systems, Toda lattices, the KdV equation, and solitons [45, 46, 51].

For the improvement of certain numerical techniques, it is often worth to construct special extrapolation procedures built on the analysis of the process to be accelerated (that is to construct extrapolations methods whose sequences in the kernel mimic as closely as possible the exact behavior of the sequence to be accelerated). For example, this was the methodology recently followed for Tikhonov regularization techniques [18], estimations of the error for systems of linear equations [12], treatment of the Gibbs phenomenon in Fourier and other series [14], and ranking in web search [35, 19, 16, 17].

Acknowledgments: I would like to thank Prof. Douglas Rogers and Prof. Haakon Waadeland for providing me informations on the life of W. Romberg. I am grateful to Prof. Peter C. Fenton for biographical informations on R.J.T. Bell and D.J. Richards. I am indebted to Michela Redivo-Zaglia for her efficient help in web searching. Finally, I acknowledge the clarifying comments of the referee (in particular her/his remarks on the interpolation/extrapolation interpretation of Aitken's process, and on Bierens de Haan, van Ceulen, van Roomen, and Seki) which greatly helped to improve parts of the paper.

References

1. A.C. Aitken, On Bernoulli's numerical solution of algebraic equations, *Proc. R. Soc. Edinb.*, **46** (1926) 289–305.
2. A.C. Aitken, Studies in practical mathematics. II. The evaluation of the latent roots and latent vectors of a matrix, *Proc. R. Soc. Edinb.*, **57** (1937) 269–304.
3. A.C. Aitken, *Determinants and Matrices*, Oliver and Boyd, Edinburgh and London, 1939.
4. A.C. Aitken, *Gallipoli to the Somme. Recollections of a New Zealand Infantryman*, Oxford University Press, London, 1963.
5. A.C. Aitken, *To Catch the Spirit. The Memoir of A.C. Aitken with a Biographical Introduction by P.C. Fenton*, University of Otago Press, Dunedin, 1995.
6. F.L. Bauer, La méthode d'intégration numérique de Romberg, in *Colloque sur l'Analyse Numérique*, Librairie Universitaire, Louvain, 1961, pp.119–129.
7. N. Bogolyubov, N. Krylov, On Rayleigh's principle in the theory of differential equations of mathematical physics and upon Euler's method in the calculus of variation (in Russian), *Acad. Sci. Ukraine (Phys. Math.)*, **3** (1926) 3–22.
8. C. Brezinski, Application de l'ε-algorithme à la résolution des systèmes non linéaires, *C.R. Acad. Sci. Paris*, **271** A (1970) 1174–1177.
9. C. Brezinski, A general extrapolation algorithm, *Numer. Math.*, **35** (1980) 175–187.
10. C. Brezinski, *History of Continued Fractions and Padé Approximants*, Springer, Berlin, 1991.

11. C. Brezinski, Extrapolation algorithms and Padé approximations: a historical survey, *Appl. Numer. Math.*, **20** (1996) 299–318.
12. C. Brezinski, Error estimates for the solution of linear systems., *SIAM J. Sci. Comput.*, **21** (1999) 764–781.
13. C. Brezinski, Convergence acceleration during the 20th century, *J. Comput. Appl. Math.*, **122** (2000) 1–21.
14. C. Brezinski, Extrapolation algorithms for filtering series of functions, and treating the Gibbs phenomenon, *Numer. Algorithms*, **36** (2004) 309–329.
15. C. Brezinski, M. Redivo-Zaglia, *Extrapolation Methods. Theory and Practice*, North-Holland, Amsterdam, 1991.
16. C. Brezinski, M. Redivo-Zaglia, The PageRank vector: properties, computation, approximation, and acceleration, *SIAM J. Matrix Anal. Appl.*, **28** (2006) 551–575.
17. C. Brezinski, M. Redivo-Zaglia, Rational extrapolation for the PageRank vector, *Math. Comp.*, **77** (2008) 1585–1598.
18. C. Brezinski, M. Redivo-Zaglia, G. Rodriguez, S. Seatzu, Extrapolation techniques for ill–conditioned linear systems, *Numer. Math.*, **81** (1998) 1–29.
19. C. Brezinski, M. Redivo-Zaglia, S. Serra-Capizzano, Extrapolation methods for PageRank computations, *C.R. Acad. Sci. Paris, Sér. I*, **340** (2005) 393–397.
20. E. Caliceti, M. Meyer-Hermann, P. Ribeca, A. Surzhykov, U.D. Jentschura, From useful algorithms for slowly convergent series to physical predictions based on divergent perturbative expansions, *Pysics Reports*, **446**, Nos. 1–3, 2007.
21. J.-L. Chabert et al., *A History of Algorithms. From the Pebble to the Microchip*, Springer, Berlin, 1999.
22. L. Collatz, *Numerische Behandlung von Differentialgleichungen*, Springer–Verlag, Berlin, 1951.
23. M. Crouzeix, A. Mignot, *Analyse Numérique des Équations Différentielles*, 2nd ed., Masson, Paris, 1989.
24. J.P. Delahaye, *Sequence Transformations*, Springer–Verlag, Berlin, 1988.
25. J.P. Delahaye, B. Germain-Bonne, Résultats négatifs en accélération de la convergence, *Numer. Math.*, **35** (1980) 443–457.
26. J. Dutka, Richardson extrapolation and Romberg integration, *Historia Math.*, **11** (1984) 3–21.
27. E. Fürstenau, *Darstellung der reellen Würzeln algebraischer Gleichungen durch Determinanten der Coefficienten*, Marburg, 1860.
28. T. Håvie, On a modification of Romberg's algorithm, *BIT*, **6** (1966) 24–30.
29. T. Håvie, Generalized Neville type extrapolation schemes, *BIT*, **19** (1979) 204–213.
30. A. Hirayama, K. Shimodaira, H. Hirose, *Takakazu Seki's Collected Works Edited with Explanations*, Osaka Kyoiku Tosho, Osaka, 1974.
31. H. Holme, Beitrag zur Berechnung des effektiven Zinsfusses bei Anleihen, Skand. *Aktuarietidskr.*, **15** (1932) 225–250.
32. J.C.R. Hunt, Lewis Fry Richardson and his contributions to mathematics, meteorology, and models of conflicts, *Annu. Rev. Fluid Mech.*, **30** (1998) xiii–xxxvi.
33. C. Huygens, *De Circuli Magnitudine Inventa*, Johannem et Danielem Elzevier, Amsterdam, 1654 and *Œuvres*, Société Hollandaise des Sciences, Martinus Nijhoff, 1910, vol. 12, pp. 91–181.
34. D.C. Joyce, Survey of extrapolation processes in numerical analysis, *SIAM Rev.*, **13** (1971) 435–490.
35. S.D. Kamvar, T.H. Haveliwala, C.D. Manning, G.H. Golub, Extrapolations methods for accelerating PageRank computations, in *Proceedings of the 12th International World Wide Web Conference*, ACM Press, New York, 2003, pp. 261-270.

36. K. Kommerell, *Das Grenzgebiet der Elementaren und Höheren Mathematik*, Verlag Köhler, Leipzig, 1936.
37. P.J. Laurent, Un théorème de convergence pour le procédé d'extrapolation de Richardson, *C. R. Acad. Sci. Paris*, **256** (1963) 1435–1437.
38. C. Maclaurin, *Treatise of Fluxions*, Ruddimans, Edinburgh, 1742.
39. G.I. Marchuk, V.V. Shaidurov, *Difference Methods and their Extrapolations*, Springer–Verlag, New York, 1983.
40. J.C. Maxwell, *A Treatise on Electricity and Magnetism*, Oxford University Press, Oxford, 1873.
41. B. Meidell, Betrachtungen über den effektiven Zinsfuss bei Anleihen, *Skand. Aktuarietidskr.*, **15** (1932) 159–174.
42. G. Meinardus, G.D. Taylor, Lower estimates for the error of the best uniform approximation, *J. Approx. Theory*, **16** (1976) 150–161.
43. R.M. Milne, Extension of Huygens' approximation to a circular arc, *Math. Gaz.*, **2** (1903) 309–311.
44. H. Naegelsbach, Studien zu Fürstenau's neuer Methode der Darstellung und Berechnung der Wurzeln algebraischer Gleichungen durch Determinanten der Coefficienten, *Arch. d. Math. u. Phys.*, **59** (1876) 147–192; **61** (1877) 19–85.
45. A. Nagai, J. Satsuma, Discrete soliton equations and convergence acceleration algorithms, *Phys. Letters A*, **209** (1995) 305–312.
46. A. Nagai, T. Tokihiro, J. Satsuma, The Toda molecule equation and the ε–algorithm, *Math. Comp.*, **67** (1998) 1565–1575.
47. N.E. Nørdlund, Johan Frederik Steffensen in memoriam, *Nordisk Matematisk Tidsskrift*, **10** (1962) 105–107.
48. T.H. O'Beirne, On linear iterative processes and on methods of improving the convergence of certain types of iterated sequences, Technical report, Torpedo Experimental Establishment, Greenock, May 1947.
49. M.E. Ogborn, Johan Frederik Steffensen, *J. Inst. Actuaries*, **88** (1962) 251–252.
50. O. Østerby, Archimedes, Huygens, Richardson and Romberg, unpublished manuscript; see www.daimi.au.dk/~oleby/notes/arch.ps
51. V. Papageorgiou, B. Grammaticos, A. Ramani, Integrable lattices and convergence acceleration algorithms, *Phys. Letters, A*, **179** (1993) 111–115.
52. L.F. Richardson, The approximate arithmetical solution by finite difference of physical problems involving differential equations, with an application to the stress in a masonry dam, *Philos. Trans. Roy. Soc. London, ser. A*, **210** (1910) 307–357.
53. L.F. Richardson, *Weather Prediction by Numerical Process*, Cambridge University Press, Cambridge, 1922; second edition, 2007.
54. L.F. Richardson, *Mathematical Psychology of War*, Hunt, Oxford, 1919.
55. L.F. Richardson, The deferred approach to the limit. Part I: Single lattice, *Philos. Trans. Roy. Soc. London, ser. A*, **226** (1927) 299–349.
56. L.F. Richardson, A purification method for computing the latent columns of numerical matrices and some integrals of differential equations, *Philos. Trans. Roy. Soc. London, Ser. A*, **242** (1950) 439–491.
57. L.F. Richardson, *Statistics of Deadly Quarrels*, Boxwood Press, Pacific Grove, 1960.
58. L.F. Richardson, The problem of contiguity: an appendix of statisticfs of deadly quarrels, *Gen. Syst. Yearb.*, **6** (1961) 139–187.
59. W. Romberg, Vereinfachte numerische Integration, *Kgl. Norske Vid. Selsk. Forsk.*, **28** (1955) 30–36.
60. H. Rutishauser, *Der Quotienten-Differenzen Algorithmus*, Birkhäuser-Verlag, Basel, 1957.

61. J.F. Saigey, *Problèmes d'Arithmétique et Exercices de Calcul du Second Degré avec les Solutions Raisonnées*, Hachette, Paris, 1859.

62. M.G. Salvadori, M.L. Baron, *Numerical Methods in Engineering*, Prentice–Hall, Englewood Cliffs, 1952.

63. R.J. Schmidt, On the numerical solution of linear simultaneous equations by an iterative method, *Phil. Mag.*, **7** (1941) 369–383.

64. C. Schneider, Vereinfachte Rekursionen zur Richardson–Extrapolation in Spezialfällen, *Numer. Math.*, **24** (1975) 177–184.

65. D. Shanks, An analogy between transient and mathematical sequences and some nonlinear sequence-to-sequence transforms suggested by it. Part I, Memorandum 9994, Naval Ordnance Laboratory, White Oak, July 1949.

66. D. Shanks, Non linear transformations of divergent and slowly convergent sequences, *J. Math. and Phys.*, **34** (1955) 1–42.

67. D. Shanks, J.W. Wrench, Jr., Calculation of π to 100.000 decimals, *Math. Comp.*, **16** (1962) 76–99.

68. D. Shanks, *Solved and Unsolved Problems in Number Theory*, Spartan Books, Washington, 1962.

69. W.F. Sheppard, Central difference formulæ, *Proc. London Math. Soc.*, **31** (1899) 449–488.

70. W.F. Sheppard, Some quadrature formulas, *Proc. London Math. Soc.*, **32** (1900) 258–277.

71. A. Sidi, *Practical Extrapolation Methods: Theory and Applications*, Cambridge University Press, Cambridge, 2003.

72. S.B. Smith, *The Great Mental Calculators. The Psychology, Methods, and Lives of Calculating Prodigies Past and Present*, Columbia University Press, New York, 1983.

73. J.F. Steffensen, *Interpolation*, Williams and Wilkins, Baltimore, 1927. Reprinted by Chelsea, New York, 1950.

74. J.F. Steffensen, Remarks on iteration, *Skand. Aktuarietidskr.*, **16** (1933) 64–72.

75. E. Stiefel, Altes und neues über numerische Quadratur, *Z. Angew. Math. Mech*, **41** (1961) 408–413.

76. G. Walz, The history of extrapolation methods in numerical analysis, Report Nr. 130, Universität Mannheim, Fakultät für Mathematik und Informatik, 1991.

77. G. Walz, *Asymptotics and Extrapolation*, Akademie Verlag, Berlin, 1996.

78. E.J. Weniger, Nonlinear sequence transformations for the acceleration of convergence and the summation of divergent series, *Comput. Physics Reports*, **10** (1989) 189–371.

79. H.C. Williams, Daniel Shanks (1917-1996), *Math. Comp.*, **66** (1997) 929–934 and *Notices of the AMS*, **44** (1997) 813–816.

80. J. Wimp, *Sequence Transformations and their Applications*, Academic Press, New York, 1981.

81. P. Wynn, On a device for computing the $e_m(S_n)$ transformation, *MTAC*, **10** (1956) 91–96.

82. P. Wynn, Acceleration techniques for iterated vector and matrix problems, *Math. Comp.*, **16** (1962) 301–322.

83. P. Wynn, Upon systems of recursions which obtain among the quotients of the Padé table, *Numer. Math.*, **8** (1966) 264–269.

Very basic multidimensional extrapolation quadrature*

James N. Lyness

Mathematics and Computer Science Division
Argonne National Laboratory
9700 South Cass Avenue
Argonne, IL 60439 USA
and
School of Mathematics
University of New South Wales
Sidney NSW 2052, Australia
lyness@mcs.anl.gov

Abstract. Extrapolation quadrature comprises a branch of numerical quadrature and cubature that developed from the (early nineteenth century) Euler-Maclaurin asymptotic expansion by intensive application of the idea (1927) of Richardson's deferred approach to the limit.

At present, extrapolation quadrature provides an elegant, well-rounded theory for a significant and well defined class of quadrature problems. This class comprises integration over an N-dimensional simplex and an N-dimensional parallelepiped, and so, by extension, over all polyhedra. It may be applied to regular integrand functions, and to integrand functions having algebraic or logarithmic singularities at vertices.

Within this class of problem, when $N > 1$ and the integrand is singular, polynomial (Gaussian) quadrature is extraordinarily cumbersome to apply, while extrapolation is simple and cost-effective. On the other hand, in one dimension and for N-dimensional regular integrands, extrapolation remains simple and cost-effective, but polynomial (Gaussian) quadrature is significantly more cost-effective.

This note is devoted to clarifying this situation. It is concerned with the underlying expansions on which extrapolation is based; and at the end, pinpointing the difference in approach to that of polynomial approximation which leads to this dichotomy.

1 Introduction

1.1 Software

This paper concerns the development of extrapolation quadrature to provide the theoretical justification for a simple algorithm for integrating certain N-dimensional singular integrands over an N-dimensional polyhedron. The resulting

---
* The author was supported by the Mathematical, Information, and Computational Sciences Division subprogram of the Office of Advanced Scientific Computing Research, Office of Science, U.S. Department of Energy, under Contract DE-AC02-06CH11357.

procedure is simple to use. The algorithm requires only two items of numerical software. These are:

(1) A standard linear equation solver. This requires input s, an $s \times s$ matrix V and an s-vector b. It returns an s-vector $x = V^{-1}b$ and usually an estimate K of the condition number of V.

(2) Subprograms for evaluating and N-dimensional m-panel rule (1.7) over C, a specified parallelepiped; or for evaluating a closely related rule (1.9) over T, a specified simplex. These implement quadrature rules which we denote by $Q^{[m]}(C)f$ or by $Q^{[m]}(T)f$ respectively. Naturally, these require input parameters which define the integration region, the positive integer panel number m; and they require access to a user-provided subroutine that is used to evaluate the integrand function f at any point within the integration region.

With these software items, one can implement all the formulas suggested in the rest of this paper for integrating large classes of singular integrands over N-dimensional polyhedra.

1.2 N-dimensional quadrature rules

We denote by $\square$ the unit cube $[0,1]^N$. The rule $Q^{[m]}(\square)f$ is simply the N-product m-panel mid-point rule. Thus

$$Q^{[m]}(\square)f = m^{-N} \sum_{j_i \in [1,m]} f(\mathbf{t_j}/m) \tag{1.1}$$

with

$$\mathbf{t_j} = ((2j_1 - 1), (2j_2 - 1), \ldots, (2j_N - 1))/2. \tag{1.2}$$

We denote by $\triangle$ the standard N-dimensional simplex obtained from $\square$ by retaining only the corner nearest the origin in which $x_1 + x_2 + \cdots + x_N \leq 1$. The rule $Q^{[m]}(\triangle)f$ is obtained from $Q^{[m]}(\square)f$ in an analogous way. One simply omits points outside the simplex, and assigns to points on the boundary an appropriate weight factor. Thus

$$Q^{[m]}(\triangle)f = m^{-N} \sum_{j_i \in [1,m]} f(\mathbf{t_j}/m)\theta(\mathbf{t_j}/m), \tag{1.3}$$

where $\mathbf{t_j}$ is as given above and $\theta(\mathbf{x}) = 1$ or 0, according as $\mathbf{x}$ is strictly inside $\triangle$, or is strictly outside $\triangle$. When $\mathbf{x}$ is on a boundary, $\theta(\mathbf{x})$ is a rational fraction representing the proportion of a spherical neighborhood of $\mathbf{x}$ lying inside $\triangle$. When N is odd, there are no boundary nodes. Minor modifications of this definition are available for even $N \geq 4$.

These rules for the standard regions may be extended to more general regions of the same type by means of an affine transformation. The rest of this section is devoted to describing this in detail. Let $\mathbf{e}_j$ be the jth unit N-vector $(0,0,\ldots,0,1,0,\ldots,0)$. The $N+1$ vertices of $\triangle$ are at $\mathbf{0}$ and $\mathbf{e}_j$, $j = 1, 2, \ldots, N$;

and the 2^N vertices of $\square$ are at $\sum_{j=1}^{N} \lambda_j \mathbf{e}_j$ with each of the 2^N possibilities obtained from $(\lambda_1, \lambda_2, \ldots, \lambda_N)$ with each λ_j taking the values 0 or 1.

We treat a simplex T having vertices at $\mathbf{0}$ and $\mathbf{a}_j$, $j = 1, 2, \ldots, N$; and a parallelepiped C having vertices at $\sum_{j=1}^{N} \lambda_j \mathbf{a}_j$. These may be obtained from the standard regions $\triangle$ and $\square$ using an *affine transformation* M. Thus

$$\mathbf{a}_j = M\mathbf{e}_j; \quad C = M\square; \quad T = M\triangle. \tag{1.4}$$

It is evident by inspection that M takes a function $f(x)$ into another function $F(x) = f(M^{-1}x)$ yielding

$$\int_C F(x)\,dx = |M| \int_{[0,1]^N} f(x)\,dx \tag{1.5}$$

with

$$M = \det(\mathbf{a}_1, \mathbf{a}_2, \ldots, \mathbf{a}_N). \tag{1.6}$$

The same transformation may be applied to both discretisations (1.1) and (1.3) leading to the following definitions.

Definition 1.1. *An m-panel product mid-point trapezoidal rule for the parallelepiped C is*

$$Q^{[m]}(C)f = |M|m^{-N} \sum_{j_i \in [1,m]} f(\mathbf{t_j}/m) \tag{1.7}$$

with

$$\mathbf{t_j} = ((2j_1 - 1)\mathbf{a}_1 + (2j_2 - 1)\mathbf{a}_2 + \cdots + (2j_N - 1)\mathbf{a}_N)/2. \tag{1.8}$$

Definition 1.2. *An m-panel mid-point trapezoidal rule for the simplex T is*

$$Q^{[m]}(C)f = |M|m^{-N} \sum_{j_i \in [1,m]} f(\mathbf{t_j}/m))\theta(\mathbf{t_j}/m) \tag{1.9}$$

with $\mathbf{t_j}$ given by the same formula and $\theta(\mathbf{x}) = 1$ or $\theta(\mathbf{x}) = 0$ according as $\mathbf{x}$ is strictly inside T, or is strictly outside T. When $\mathbf{x}$ is on a boundary, $\theta(\mathbf{x})$ is a rational fraction representing the proportion of a spherical neighborhood of $\mathbf{x}$ lying inside T.

Note that $|M|$ is the N-volume of C. The N-volume of T is $|M|/N!$.

Looking ahead for a moment, we see that by applying the transformation in (1.5) to the quadrature error to obtain

$$Q^{[m]}(C)F - I(C)F = |M| \left(Q^{[m]}(\square)f - I(\square)f \right); \tag{1.10}$$

$$Q^{[m]}(T)F - I(T)F = |M| \left(Q^{[m]}(\triangle)f - I(\triangle)f \right). \tag{1.11}$$

These simple results allows us to obtain error expansions for the standard regions and apply them immediately to the linearly transformed regions. A similar remark is not valid for multidimensional Gaussian quadrature. Indeed, Gaussian quadrature is infeasible to the extent of being almost impossible for the problems described in section 3 but extrapolation is a viable approach. This aspect is discussed in section 5.

2   Extrapolation quadrature for regular integrands

2.1   One dimension; regular integrand

Our starting point is an almost unbearably simple example of extrapolation quadrature. Here $f(x)$ is regular in $[0, 1]$, that is $f \in C^\infty[0, 1]$. A straightforward approximation to the integral.

$$If = \int_0^1 f(x)dx, \tag{2.1}$$

is provided by the m-panel mid-point trapezoidal rule

$$Q^{[m]}f = \frac{1}{m} \sum_{j=1}^m f\left(\frac{2j-1}{2m}\right). \tag{2.2}$$

This requires m function values and, in isolation, is not in general a particularly good approximation. This rule is in fact the m-copy version of the one point mid-point rule, which we may denote by

$$Qf = Q^{[1]}f = f(1/2). \tag{2.3}$$

For larger m, the m-copy version is generally, but not always, more accurate. This dependence on m is quantified by the classical Euler-Maclaurin asymptotic expansion, which takes the following form.

Theorem 2.1. *Let $f \in C^\infty[0, 1]$. Let m, p be positive integers. Then*

$$Q^{[m]}f = If + \frac{B_2}{m^2} + \frac{B_4}{m^4} + \cdots + \frac{B_{2p}}{m^{2p}} + R_{2p+2}(m) \tag{2.4}$$

where B_j is independent of m and $R_j(m) = O(m^{-j})$.

The infinite series obtained from (2.6) is generally not convergent.

Extrapolation quadrature makes use of several approximations of this type, and combines them in a way designed to provide what one hopes may be a closer approximation to If. Suppose we have available four of these approximations, those having mesh values $m = m_1, m_2, m_3$, and m_4. Then we can rewrite the set of four equations obtained from (2.4) using these mesh values and $p = 3$ in matrix form as

$$
\begin{pmatrix} Q^{[m_1]}f \\ Q^{[m_2]}f \\ Q^{[m_3]}f \\ Q^{[m_4]}f \end{pmatrix} = \begin{pmatrix} 1 & m_1^{-2} & m_1^{-4} & m_1^{-6} \\ 1 & m_2^{-2} & m_2^{-4} & m_2^{-6} \\ 1 & m_3^{-2} & m_3^{-4} & m_3^{-6} \\ 1 & m_4^{-2} & m_4^{-4} & m_4^{-6} \end{pmatrix} \begin{pmatrix} If \\ B_2 \\ B_4 \\ B_6 \end{pmatrix} + \begin{pmatrix} R_8(m_1) \\ R_8(m_2) \\ R_8(m_3) \\ R_8(m_4) \end{pmatrix}
$$

(2.5)

and express this in matrix notation as

$$ Q = VI + R. \tag{2.6} $$

We are interested in calculating If, the first element of the vector

$$ I = V^{-1}Q - V^{-1}R. \tag{2.7} $$

We have available numerical values of the elements of Q, and the elements of the Vandermonde matrix V. Since we do not know R, in the best traditions of numerical calculation, we abandon it and calculate instead the first element of

$$ \tilde{I} = V^{-1}Q, \tag{2.8} $$

thereby introducing a discretisation error, the first element of

$$ \tilde{I} - I = V^{-1}R. \tag{2.9} $$

All that is needed to calculate the first element of $\tilde{I}$ is a set of values of m together with a corresponding set of $Q^{[m]}f$. Then we need only apply a standard linear equation solver to solve $V\tilde{I} = Q$. But remember, the connecting equation (2.4) is critical, as this justifies the entries in V, the Vandermonde matrix. In all the generalisations of this technique known to the author, the key result is the generalisation of this expansion, which is also known as the error functional. Without this theoretical result being available, no reliable progress is possible.

2.2 N-Dimensional square and simplex; regular integrand

Since extrapolation quadrature in one dimension is so straightforward, it is natural to look for other contexts in which the same sort of extrapolation is available. The first is really obvious: numerical quadrature over a hypercube $[0,1]^N$. The rule $Q^{[m]}(\square)f$ defined in section 1 is an N-dimensional Cartesian product of the one-dimensional m-panel mid-point trapezoidal rule. Since the hypercube is a product region, a relatively simple derivation reveals an expansion of precisely the same form as (2.4) above. Naturally, the coefficients B_j in the new expansion are different from those in (2.4) but we recall they play no major role in the subsequent calculation. The elements of the Vandermonde matrix coincide precisely with those in the one dimensional case. Again the approximation $\widetilde{If}$ is the first element of $\tilde{I} = V^{-1}Q$.

Another context is in quadrature over a simplex.

Theorem 2.2. [3, **Theorem 2.2**] *Let* $f \in C^{\infty}(\triangle)$ *where* $\triangle$ *is the standard* N-*dimensional simplex. Let* m, p *be positive integers. Then*

$$Q^{[m]}(\triangle)f \sim I(\triangle)f + \frac{B_2}{m^2} + \frac{B_4}{m^4} + \cdots + \frac{B_{2p}}{m^{2p}} + R_{2p+2}(m) \qquad (2.10)$$

where B_j *is independent of* m.

I know of only two proofs of this expansion. Both are cumbersome, involving several journal pages. However the result is important. It provides the justification for extending the technique to simplices.

Note that, while these expansions are derived and stated in terms of standard hypercubes and simplices, the theory can readily be extended to general parallelepipeds and simplices by means of an affine transformation.

These are the situations of which I am aware in which extrapolation can be based on the Euler-Maclaurin expansion without any modification. Note that in these situations the integrand is regular.

However, it seems that the product Gauss-Legendre rules for the hypercube, and the conical product rule for the simplex are respectively more cost effective but by a slightly smaller relative margin as the dimension becomes higher.

3   Extrapolation quadrature for some N-dimensional algebraic singularities

3.1   An N-dimensional example

We now proceed to integrands which have a singularity of a particular type at a vertex. To fix ideas we treat a simple case,

$$If = \int_{[0,1]^N} r^{\alpha} d^N x, \quad \text{with} \quad r = \sqrt{x_1^2 + x_2^2 + \cdots + x_N^2}. \qquad (3.1)$$

Here $f(x) = r^{\alpha}$ and we need $\alpha + N > 0$ for the integral to exist in the conventional sense. When α is an even nonnegative integer, $f \in C^{\infty}[0, 1]$, in which case the expansion (2.4) applies. Otherwise expansion (2.4) is not valid because the integrand is not C^{∞} over the closed integration region.

In this another expansion may be used in place of (2.4), namely

$$Q^{[m]}f \sim If + \frac{A_{\alpha+N}}{m^{\alpha+N}} + \frac{B_2}{m^2} + \frac{B_4}{m^4} + \cdots \quad \text{when } \alpha + N \text{ is not an even integer} \quad (3.2)$$

$$Q^{[m]}f \sim If + \frac{C_{\alpha+N} \log m}{m^{\alpha+N}} + \frac{B_2}{m^2} + \frac{B_4}{m^4} + \cdots \quad \text{when } \alpha + N \text{ is an even integer.} \qquad (3.3)$$

This pair of expansions comprise an example of Theorem 3.1 below. Note particularly that the coefficients denoted here by B_q are quite different from those which occur in previous theorems. And there is one additional term. Otherwise, the form is the same, and it can be exploited in just the same way. In the first case (that is when $\alpha + N$ is not an even integer) our equation $Q = V\tilde{I}$ is:

$$
\begin{pmatrix} Q^{[m_1]}f \\ Q^{[m_2]}f \\ Q^{[m_3]}f \\ Q^{[m_4]}f \end{pmatrix} = \begin{pmatrix} 1 & m_1^{-(\alpha+N)} & m_1^{-2} & m_1^{-4} \\ 1 & m_2^{-(\alpha+N)} & m_2^{-2} & m_2^{-4} \\ 1 & m_3^{-(\alpha+N)} & m_3^{-2} & m_3^{-4} \\ 1 & m_4^{-(\alpha+N)} & m_4^{-2} & m_4^{-4} \end{pmatrix} \begin{pmatrix} \widetilde{If} \\ \tilde{A}_{\alpha+N} \\ \tilde{B}_2 \\ \tilde{B}_4 \end{pmatrix}.
$$

In the second case, the one in which $\alpha + N$ is an even integer, the equation is

$$
\begin{pmatrix} Q^{[m_1]}f \\ Q^{[m_2]}f \\ Q^{[m_3]}f \\ Q^{[m_4]}f \end{pmatrix} = \begin{pmatrix} 1 & m_1^{-(\alpha+N)}\log m_1 & m_1^{-2} & m_1^{-4} \\ 1 & m_2^{-(\alpha+N)}\log m_2 & m_2^{-2} & m_2^{-4} \\ 1 & m_3^{-(\alpha+N)}\log m_3 & m_3^{-2} & m_3^{-4} \\ 1 & m_4^{-(\alpha+N)}\log m_4 & m_4^{-2} & m_4^{-4} \end{pmatrix} \begin{pmatrix} \widetilde{If} \\ \tilde{C}_{\alpha+N} \\ \tilde{B}_2 \\ \tilde{B}_4 \end{pmatrix}
$$

One may proceed, exactly as before, to solve the appropriate set of equations using, perhaps the same linear equation solver as was mentioned in section 1. In this example, the procedure is extraordinarily straightforward; but it does depend critically on knowing the correct asymptotic expansion for $Q^{[m]}f - If$, in order to construct the matrix V. That is, we must use (3.2) or (3.3) and *not* (2.4). We now proceed to state the underlying theory.

3.2   Homogeneous type singularities

The integrand (3.1) of the example is one of a wide class of integrands for which an expansion of form (3.2) or (3.3) exists. When the integrand function has a singular behavior anywhere in $[0,1]^N$, (2.4) is generally not valid. However, an expansion is known for integrand functions that are homogeneous of specified degree about the origin and are $C^\infty[0,1]^N \setminus \{\mathbf{0}\}$.

Definition 3.1. $f(\mathbf{x})$ *is **homogeneous** about the origin of degree λ if $f(k\mathbf{x}) = k^\lambda f(\mathbf{x})$ for all $k > 0$ and $|\mathbf{x}| > 0$.*

For example, in two dimensions, let $A \neq 0$ and B be constants. Then functions such as

$$
(Ax^2 + By^2)^{\lambda/2}, \quad (Ax + By)^\lambda, \quad (xy^2)^{\lambda/3}, \tag{3.4}
$$

are homogeneous of degree λ about the origin.

Theorem 3.1. [1] *Let* $\gamma > N$; *let* $f(\mathbf{x})$ *be* **homogeneous** *of degree* γ *and be* $C^\infty[0,1]^N \setminus \{\mathbf{0}\}$. *Then*

$$Q_N^{[m]} f \sim If + \frac{C_{\gamma+N} \log m + A_{\gamma+N}}{m^{\gamma+N}} + \sum_{s=1} \frac{B_{2s}}{m^2}, \tag{3.5}$$

where the coefficients are independent of m *and* $C_j = 0$, *unless* j *is an even integer.*

The only proof of this known to the author, requires several journal pages. In the example of the previous section we have applied this theorem to r^α, which is homogeneous of degree α. Incidentally, (3.5) is completely equivalent to the pair (3.2), (3.3).

Definition 3.2. $f(\mathbf{x})$ *is termed* **pseudohomogeneous** *of degree* λ *about the origin if it may be expressed in the form* $f(\mathbf{x}) = f_\lambda(\mathbf{x})h(\mathbf{x})$, *where* $f_\lambda(\mathbf{x})$ *is homogeneous of degree* λ *about the origin and* $h(\mathbf{x})$ *is analytic in a neighborhood of the origin.*

A regular function h is pseudohomogeneous of integer degree 0. It may also be of higher integer degree. In two dimensions the function (using conventional polar coordinate notation with g, h and Θ regular)

$$r^\alpha g(r)h(x,y)\Theta(\theta)$$

is pseudohomogeneous of degree α. In the sequel, we shall omit the phrase "about the origin". We note that f, a pseudohomogeneous function of degree λ can be expressed as

$$f = f_\lambda + f_{\lambda+1} + f_{\lambda+2} + \cdots + f_{\lambda+p} + G,$$

where G has a higher degree of continuity than $f_{\lambda+p}$. Developing this approach leads to the following relatively straightforward extension of the previous theorem.

Theorem 3.2. *Let* $\gamma > -N$; *let* $f(\mathbf{x})$ *be* **pseudohomogeneous** *of degree* γ *and be* $C^\infty[0,1]^N \setminus \{\mathbf{0}\}$. *Then,*

$$Q_N^{[m]} f \sim If + \sum_{t=0} \frac{C_{\gamma+N+t} \log m + A_{\gamma+N+t}}{m^{\gamma+N+t}} + \sum_{s=1} \frac{B_{2s}}{m^{2s}}, \tag{3.6}$$

where the coefficients are independent of m *and* $C_j = 0$, *unless* j *is an even integer.*

The theorems of this section have been presented in the context in which they first appeared; that is they specify the unit cube $[0,1]^N$ as the domain of integration. In fact, they are far more general. and may be applied word for word to parallelepipeds. The reason is that an **affine transformation** M takes a (pseudo)homogeneous function into another (pseudo)homogeneous function of the same degree. If we set $F(x) = f(M^{-1}x)$ and $C = M[0,1]^N$, we find

$$\int_C F(x)dx = |M| \int_{[0,1]^N} f(x)dx.$$

The same relation applies to the discretisation $Q^{[m]}$ of the integral, and so to their difference, giving

$$Q^{[m]}(C)F - I(C)F = |M|(Q^{[m]}(\square)f - I(\square)f).$$

The upshot is the following.

Theorem 3.3. *Theorems 3.1 and 3.2 above remain valid as written when $Q^{[m]}f$ and If are redefined as a rule and the corresponding integral over a parallelepiped C.*

In fact, the scope of these results is even wider. They may be applied to simplices too. Following is an outline of some of the proof.

One may synthesise the N-dimensional cube C into two parts, namely the N-dimensional simplex T together with a complementary section $(C - T)$. In all the results of this and earlier sections, the integrand function has no singularity in the region $C - T$. The rule $Q^{[m]}$ has been defined in such a way that (in an obvious notation)

$$Q^{[m]}(T)f = Q^{[m]}(C)f - Q^{[m]}(C - T)f.$$

When there is singular behaviour at the origin, but not elsewhere in C, the final term involves directly no singular behavior. While certainly not immediate, by exploiting Theorem 2.2 it is not too difficult to show that the expansion corresponding to this final term only involves terms B_{2s}/m^{2s}. The form of the expansion of the first term on the right is unaltered by these additional terms. This leads to the following theorem.

Theorem 3.4. *Theorems 3.1 and 3.2 above remain valid as written when $Q^{[m]}f$ and If are redefined as a rule and the corresponding integral over a simplex T.*

4   Choice of mesh sequence

In previous sections we have made no detailed suggestions about the choice of meshes $m_1, m_2, \ldots$. An instinctive first choice might be to reduce the cost by choosing these to be as small as feasible. This would be relatively easy to justify if the calculation were to be carried out in infinite precision arithmetic.

However, infinite precision arithmetic is rarely available. A potential user would be right to be concerned that his matrix V may be ill-conditioned, introducing non-trivial amplification of the inherent noise level in the elements of Q. This situation is discussed in some detail in [2].

Many linear equation solvers provide an estimate of the condition number K of the matrix V. While it is provided automatically, it may be unduly pessimistic for our purposes. We are interested in a condition number, $\overline{K}$ corresponding to the *first element* $\widetilde{If}$ of the vector $\tilde{I}$. Naturally, this cannot exceed K, but in the type of matrices we are dealing with here, it can be significantly smaller than K.

With certain minor restrictions, this condition number may be calculated at the same time as or before $\widetilde{If}$ is calculated. $\overline{K}$ coincides with the magnitude of the first element of $V^{-1}U$ with $U = (1, -1, 1, -1, \ldots)^T$. The restrictions are that the terms in the asymptotic expansion should be treated in their natural order, and that the final included term should be an inverse power of m (and not a term involving $\log m$).

5   Gaussian formulas for singular integrands

In section 2.1, we completed our treatment of one dimensional regular integrands. At the end of that section I remarked that, up to that point, a Gaussian rule approach seemed to be more cost-effective than extrapolation. However, for the integrands of subsequent sections (singular integrands in more than one dimension), this advantage seems to disappear completely.

We outline briefly the very well known Gaussian rule approach for handling a two or more dimensional integral whose integrand is of the form $f(x) = w(x)h(x)$ where $w(x)$ contains the singularity and $h(x)$ is a regular function, which might be readily approximated by a polynomial. The quadrature rule is of the form $Qf = \sum_i w_i h(x_i)$ and is of polynomial degree d when it has the property that $Qf = If$ for all polynomials $h(x)$ of degree d or less. In the present context the weight function may be identified as $f_\lambda(x)$ appearing in Definition 3.2. A major difference between Gaussian rules and rules based on extrapolation is that one uses function values of $h(x)$ and the other uses function values of $f(x)$.

In multidimensional quadrature of singular integrands, it often appears difficult to locate a Gaussian rule. There are two related reasons. The first is that a major calculation is required to construct Gaussian formulas. The second is that, once constructed, they are quite inflexible; perhaps parochial would be a better term.

Let us deal with the second point first. An affine transformation takes one Gaussian formula into another. But, by the same token, it also takes the weight function into another weight function. For example, suppose one has available a Gaussian formula for the standard triangle (vertices at $(0,0),(1,0),(0,1)$) with a weight function $w(x,y) = 1/r$. And suppose one required such a formula, with the same weight function, but a differently shaped triangle (for example one having vertices at $(0,0),(4,0),(0,1)$). The affine transformation which takes the first triangle into the second also alters the weight function to $w(x,y) = 1/(x^2 + 4y^2)^{1/2}$. The upshot seems to be that, unless the weight function is constant, one needs a new Gaussian formula whenever the shape of the triangle is changed. This is *not* the case when extrapolation is used, since both weight functions are homogeneous of the same order, the *procedure* for handling these different problems using extrapolation quadrature is identical.

This situation would not be so bad if it were straightforward to construct Gaussian formulas. In one dimension, in the early days, many formulas were tabulated. Later they could be obtained from subroutine libraries. More recently, programs have been constructed which handle almost any one-dimensional weight function providing weights, abscissas and even carrying out a set of integrations at the same time. This reflects to some extent the advanced state of the theory of (one dimensional) orthogonal polynomials.

The current situation in two or more dimensions is quite different. The theory of orthogonal polynomials is much more complex and has not reached the stage of supplying a sequence of formulas. Cubature rules are conventionally constructed by solving sets of (non-linear) moment equations. And, during the course of such a project, no natural sequence of rules.

In fact, if one were to attempt to construct a set of Gaussian formulas for integrands like those in section 3, possibly the first task would be to calculate the moments using the extrapolation methods of those sections. While one cannot completely rule out any particular method, it appears that in many cases involving N-dimensional singularities, the Gaussian approach is simply not competitive.

6   Concluding remarks

This paper is a brief no-frills account of the theory on which the technique now known as Extrapolation Quadrature is based. One of its purposes is to assemble the theory in a way that it can be immediately comprehended and used. I believe that all the theory described here was available in 1977 (over thirty years ago.)

My intention is that a reader with no prior acquaintance with extrapolation quadrature, can, if he wishes, complete an accurate integration of some singular function mentioned in section 3 within a couple of hours. (This would include both reading the theory and programming the calculation.) To do so is possible because I have omitted large sections of material that woud normally be expected in an account of such a wide attractive theory.

(1) Scope. The topic is limited to integrating over simplices and parallelepipeds. The integrand may be regular, or may have a singularity of an algebraic nature (see Definition 3.1) at a vertex. When there is a singularity, in higher dimensions, the Gaussian approach is extraordinarily difficult to implement and extrapolation is straightforward. (See section 5.)

(2) The numerical sofware required is simple. One requires only a linear equation solver and a numerical quadrature routine for evaluating trapezoidal type quadrature rule approximations.

(3) One needs to choose the error expansion appropriate to the singularity. These occur throughout sections 2 and 3 as theorems. Only the structure is needed and this is almost simple enough to commit to memory.

(4) An unfortunate aspect of this theory lies in the proofs of the two key error expansions (theorems 2.2 and 3.1). The proofs known to the author appear to be long, involved and nontransparent.

A subsidiary purpose of this paper is to draw attention to item 4. This is a blemish on a fine theory. Even a casual reader will see that these theorems are central to the theory. They deserve elegant proofs! I am hoping that some reader of this paper may provide a set of shorter, better proofs.

References

1. Lyness, J.N., An error functional expansion for N-dimensional quadrature with an integrand function singular at a point. *Math. Comp.* **30**, no. 133, (1976), pp. 1–23
2. Lyness, J.N., Applications of extrapolation techniques to multidimensional quadrature of some integrand functions with a singularity. *J. Computational Phys.* **20**, no. 3, (1976), pp. 346–364.
3. Lyness, J.N.; Puri, K.K., The Euler-Maclaurin expansion for the simplex. *Math. Comp.* **27** (1973), pp. 273–293.

Numerical methods for ordinary differential equations: early days

John C. Butcher

Department of Mathematics,
The University of Auckland
Private Bag 92019
Auckland 1142, NEW ZEALAND
butcher@math.auckland.ac.nz

Abstract. A personal account of the author's introduction to numerical methods for ordinary differential equations together with his impressions of some of the notable developments in this subject and some of the people who have contributed to these developments.

1 Introduction

This note is about "early days" but what early days and whose early days? The early days of numerical analysis, specifically the numerical analysis of ordinary differential equations, were too long ago for the author to have any real connection with them. What about the early days of numerical analysis in New Zealand? It is almost laughable to think that a small isolated country like New Zealand would have any significant history in this mathematically and technically profound subject but there is indeed some history. Two of the great pioneering figures in British numerical analysis, that is L. J. Comrie [39], and A. C. Aitken [37] were New Zealanders. Amongst Comrie's many contributions was the perfecting of the table makers art, ironically at a time when the rising importance of electronic computing was about to make table making obsolete. Aitken belongs correctly in the annals of Scottish Numerical Analysis. He was an outsider who found a home in one of the centres of scientific activity of his day. Unfortunately, this is typical of New Zealanders: anything they can contribute, they contribute somewhere else. The author of this note hopes he can be regarded as an exception. Whether his contributions are important or not, they are definitely New Zealand contributions. In this note he will say something about his own early days in the context of some of the notable events and pioneering contributors to the subject.

In 1957 the author had the privilege of hearing S. Gill and A. R. Merson, speak about their work. In 1965 he met such pioneers in the subject as P. Henrici and C. W. Gear and in 1970 he met the already famous G. Dahlquist as well as two people who were also destined to become famous: E. Hairer and G. Wanner.

Thus writing about his own early days, and about the people he encountered, does not restrict the author and is quite consistent with broader historical aims.

The paper is organised as follows. In Section 2 a survey will be presented of some of the great events, great people and great ideas in the early history of numerical methods for ordinary differential equations. In Section 3 the author will say a little about his own early days and how he got into this subject.

2   Notable events, ideas and people

In 1995, regarded as the centenary of the Runge–Kutta method, a conference was held at the CWI in Amsterdam to celebrate this occasion and, subsequently G. Wanner and the author wrote an account of the history of Runge–Kutta methods as we understood it at the time [11]. A more general history of numerical methods for ordinary differential equations appeared a few years later as a millennium review [10]. In the light of these existing surveys, it seems appropriate to aim for something on a smaller scale in the present paper.

Linear multistep methods

The classical Euler method [19], in which each time step approximates the average rate of change of the solution by its value evaluated at the beginning of the step is like a numerical Tortoise. It is slow and steady and, although it might not always win the race, it usually gets there in the end. More sophisticated methods which attempt to achieve greater accuracy by making more use of past calculations (linear multistep methods) or more intensive calculations in the current step itself, are more like hares. They are fast and flighty but quite often they get distracted and don't do as well as the tortoise.

The general form of linear multistep methods for an initial value problem

$$y' = f(x, y), \qquad y(x_0) = y_0,$$

is based on evaluating approximations $y_n \approx y(x_n)$ using an expression of the form

$$y_n = \sum_{j=1}^{k} \alpha_j y_{n-k} + h \sum_{j=0}^{k} \beta_j f(x_{n-j}, y_{n-j}).$$

Note that if $\beta_0 \neq 0$, the method is implicit (in contrast to explicit methods where $\beta_0 = 0$) and the solution of non-linear algebraic equations is required to actually evaluate y_n.

The first important of these hare-like methods is the systematic collection known as Adams-Bashforth [2] methods which, together with Adams-Moulton methods [35], are the basis of modern predictor-corrector methods. In the Adams-Bashforth methods $\beta_0 = 0$ and the only non-zero values of the α_j are $\alpha_1 = 1$. The Adams-Moulton methods are more accurate and are implicit.

A typical Adams-Bashorth method, $y_n = y_{n-1} + \frac{3}{2}hf(y_{n-1}) - \frac{1}{2}hf(y_{n-2})$ is based on the quadrature formula

$$\int_{-1}^{0} \phi(x)dx \approx \frac{3}{2}\phi(-1) - \frac{1}{2}\phi(-2).$$

If, instead of numerical integration, we start from numerical differentiation

$$\phi'(0) = \frac{3}{2}\phi(0) - 2\phi(-1) + \frac{1}{2}\phi(-2)$$

we obtain the second order backward difference method

$$y_n = \frac{2}{3}hf(y_n) + \frac{4}{3}y_{n-1} - \frac{1}{3}y_{n-2}.$$

The family of "backward-difference methods", of which this is an example, have an important role in the numerical solution of stiff problems, which we will now discuss.

Stiffness

In a seminal paper, Curtiss and Hirschfelder [12] considered a special type of difficult problem characterised by unstable behaviour unless the stepsize is unreasonably small. The increasing power of computing equipment at that time had made it feasible to attempt large problems, such as those arising in the method of lines for partial differential equations, and this type of difficult behaviour is typical of such problems. The remedy for stiffness, as it became known, is either to use such reduced stepsizes as to make the use of traditional methods possible, or to use methods especially designed for stiff problems. These stiff methods are usually implicit and this makes the cost of each step very high. But, overall, the use of these methods may be cheaper than taking the large number of steps required for explicit methods to work successfully.

Runge–Kutta methods

In the search for improvements over the Euler method, there are two natural generalisations. The linear multistep idea, in which information from recently completed integration steps is exploited, is matched by a completely different idea in which the function f defining a given differential equation is evaluated more than once in each step. The first paper on this type of generalisation, by C. Runge [38], was followed by contributions from K. Heun [30] and W. Kutta [33]. In carrying the integration from $y_{n-1} \approx y(x_{n-1})$ to $y_n \approx y(x_n) = y(x_{n-1}+h)$, where h is the stepsize, a number of stages are evaluated each defined as an approximation to y evaluated at a point close to x_{n-1}. Each stage is equal to y_{n-1} plus a linear combination of "stage derivatives" evaluated from previously computed stages. The coefficients in the various linear combinations are usually written, together with other relevant information, in a tableau

$$
\begin{array}{c|c}
c & A \\
\hline
 & b^T
\end{array}
=
\begin{array}{c|ccccc}
0 & 0 \\
c_2 & a_{21} \\
c_3 & a_{31} & a_{32} \\
\vdots & \vdots & \vdots & \ddots \\
c_s & a_{s1} & a_{s2} & \cdots & a_{s,s-1} \\
\hline
 & b_1 & b_2 & \cdots & b_{s-1} & b_s.
\end{array}
$$

The entries in this tableau indicate that the stages are defined by $Y_i = y_{n-1} + h \sum_{j=1}^{i-1} a_{ij} F_j$, where the stage derivatives are defined by $F_i = f(x_{n-1} + hc_i, Y_i)$. The tableau entries further indicate that the output approximation is equal to $y_n = y_{n-1} + h \sum_{i=1}^{s} b_i F_i$. By introducing elements above and on the diagonal of A, the possibility is allowed for a Runge–Kutta method to be implicit.

Linear and non-linear stability

To model stiff behaviour, Dahlquist [14], introduced the famous "A-stability" definition. In its more general formulation for application to a wider class of numerical methods, the definition is concerned with a linear problem $y' = qy$, where q is a complex number. We will write $z = hq$ and it is observed that for linear multistep methods, Runge–Kutta methods, and a wide range of related methods, the value of each step value y_n becomes the solution to a difference equation whose coefficients are functions of z. The set of z values for which this difference equation has only bounded solutions is known as the "stability region". A method is A-stable if all points in the left-half complex plane belong to the stability region. Although A-stability is not possible for high order linear multistep methods, it is available for high order implicit Runge–Kutta methods. Many attempts have been made to weaken this property in an appropriate way; the most important of these is A(α)-stability. This property holds for many methods for which strict A-stability, which is the same as A($\pi/2$)-stability, does not hold and these methods are effective for the solution of many stiff problems. It is also possible to strengthen the property and get even better performance for some stiff problems. It is very difficult to predict stable behaviour for non-linear problems and a new approach for doing this was discovered by Dahlquist in 1975 [15] along with the invention of "one-leg methods". The ideas associated with non-linear stability were extended first to Runge–Kutta methods [9] and later to general linear methods [4].

The starting point for all these non-linear stability investigations, is the a test equation

$$y'(x) = f(x, y(x)), \tag{2.1}$$

where

$$\langle Y - Z, f(X, Y) - f(X, Z) \rangle \leq 0. \tag{2.2}$$

From (2.2) it follows that, if $y(x)$ and $z(x)$ are two solutions to (2.2), then $\|y(x) - z(x)\|$ is non-increasing. A Runge–Kutta method is BN-stable if

$$\|y_n - z_n\| \leq \|y_{n-1} - z_{n-1}\| \tag{2.3}$$

and this is equivalent, for irreducible methods, to the requirements that $b_i > 0$ for $i = 1, 2, \ldots, s$ and that the symmetric matrix

$$M = \mathrm{diag}(b)A + A^T \mathrm{diag}(b) - bb^T,$$

is positive semi-definite.

The relationship between algebraic stability, as this type of non-linear stability is named for the all-embracing general linear case, is complicated by the fact that more than a single piece of information is passed between steps. If y_n in (2.3) is interpreted in this multivalue approximation sense, then $\| \cdot \|$ has to be generalised to reflect this more general setting. Dahlquist had already introduced the G matrix in the one-leg and linear multistep cases and this remained as a key component in the general linear formulation of algebraic stability.

Order barriers

A linear k-step method, in its most general form has $2k + 1$ parameters to be chosen and it might seem that order $2k$ is possible. This is true but the unique method of this order is unstable. What the strict order limitations are was solved by Dahlquist. His result states that for stable linear k-step methods the order cannot exceed $k + 2$ and if k is odd, the order cannot exceed $k + 1$.

In contrast to this "first Dahlquist barrier" there is a second barrier relating A-stability to order. This states that, for a linear multistep method the maximum order consistent with A-stability is only two. This result can be generalised to a what was formerly known as the Daniel-Moore conjecture. Although the second Dahlquist barrier was proved by Dahlquist, a new proof, which is easily generalised to the Daniel-Moore statement, was proved using order stars [40].

In the case of Runge–Kutta methods order barriers also exist. For explicit methods with s stages, the maximum possible order is s, up to $s = 4$ but for $s > 4$, at least $s+1$ stages are required [7]. For order $p = 8$ for example, 11 stages are required. In the case of implicit methods for which the stability function is a rational function of degrees n (numerator) and d (denominator), an order $p = n + d$ is possible. These high order "Padé approximations" are consistent with A-stability, if and only if $d - n \in \{0, 1, 2\}$. The final steps in this result were proved using order stars [40].

3   First contacts with numerical analysis

As an undergraduate mathematics student I didn't know what numerical analysis was but I knew it was an inferior subject because my Professor, H. G. Forder, said that it was. How could anyone take an interest in anything so devoid of intellectual content he asked. When I failed to get a scholarship to study mathematics in England after going as far as I could in New Zealand, my life seemed to be over. However, my mother saw an advertisement which led to my getting a research studentship at the University of Sydney. I could work on any subject as long as it used the SILLIAC computer [17]. I was persuaded to work on simulation of cosmic ray showers and I had several papers in this area in collaboration with Professor H. Messel and others. One of my early attempts at scientific computation was the development of an efficient way to calculate inverse Laplace transforms. I presented this at a conference in South Australia at which a number of eminent British numerical analysts were present.

S. Gill and R. H. Merson

The most famous contribution of Stanley Gill to numerical differential equations, is his adaptation of one of Kutta's fourth order methods to economise on memory use. His algorithm [26], which has the following tableau,

$$
\begin{array}{c|cccc}
0 & 0 \\
\frac{1}{2} & \frac{1}{2} \\
\frac{1}{2} & \frac{\sqrt{2}-1}{2} & \frac{2-\sqrt{2}}{2} \\
1 & 0 & -\frac{\sqrt{2}}{2} & \frac{2-\sqrt{2}}{2} \\
\hline
& \frac{1}{6} & \frac{2-\sqrt{2}}{6} & \frac{2+\sqrt{2}}{6} & \frac{1}{6}
\end{array}
$$

was also a precursor to the compensated summation technique [31]. In his analysis of the order conditions he made use of combinations of partial derivatives which are now known as elementary differentials. Merson took this idea further and developed an operational calculus for deriving Runge–Kutta methods [34]. A particular method he promoted was quite popular for many years:

$$
\begin{array}{c|ccccc}
0 & 0 \\
\frac{1}{3} & \frac{1}{3} \\
\frac{1}{3} & \frac{1}{6} & \frac{1}{6} \\
\frac{1}{2} & \frac{1}{8} & 0 & \frac{3}{8} \\
1 & \frac{1}{2} & 0 & -\frac{3}{2} & 2 \\
\hline
& \frac{1}{6} & 0 & 0 & \frac{2}{3} & \frac{1}{6}
\end{array}
$$

The output values is fourth order and the result computed as the fifth stage, Y_5, is used for error estimating purposes. The value of Y_5 is actually a third order approximation but for linear problems it is also order 4.

In the long discussion following Merson's talk, the following contribution was made:

Dr. S. Gill, Ferranti Ltd.

I looked into this subject 5 years ago but I did not carry the subject quite as far as Mr. Merson for two reasons.

Firstly, automatic computing was new then and there did not seem to be much interest in a variety of methods for integrating differential equations. Since then it has become apparent that there are applications for a number of different processes and it is perhaps now worthwhile standardising the procedures for developing variations of the process.

The other reason is that, not being a pure mathematician, I was never quite sure of what I was talking about. It is difficult to keep a cool head and I think this would justify a solid attack by a pure mathematician to put everything on a sound basis. I did however see the one-one correspondence between the trees and the "basic operators".

At this stage of my life I was too shy to speak to either Mr. Merson or Dr. Gill but their work had a strong influence on me. I worked on questions related to Runge–Kutta methods as a hobby for several years until my first paper on the subject was eventually published in the Journal of the Australian Mathematical Society, after being rejected by Numerische Mathematik. By this time my hobby had become my career.

In 1965 I travelled to USA and met some very famous numerical analysts.

G. Forsyth and G. Golub

George Forsyth of Stanford University, was a pioneer in numerical linear algebra and the academic father and grandfather of many of the present day leaders in numerical analysis. He was also one of the visionaries who viewed Computer Science, and Computational Mathematics as an integral part of it, as a distinct academic discipline. His famous protegé, Gene Golub, became a lifelong friend and supporter for myself, and for many other people. He is credited with the rhetorical question: "Numerical ODEs: is there anything left to do?". This challenge was answered by Bill Gear in an interesting paper [24].

C. W. Gear and P. Henrici

I met both Bill Gear and Peter Henrici during my first year in USA. Bill has made many contributions to this subject but I will mention only three. The first is his invention of "hybrid methods" [20] (actually the co-inventor see also [27] and [6]) which opened the way to a wide variety of multi-stage multi-value methods. The second is his championing, and further developing, the Nordsieck adaptation of Adams and other linear multistep methods [36], [21], [22]. Finally, the DIFSUB code, which is the first general purpose variable step, variable order, differential equation solver [23].

Peter Henrici wrote an important textbook on Numerical ODEs [29] which is acknowledged to be a masterpiece of exposition. It is also significant in aiming for a high mathematical style without deviating from its core aim of providing analyses and critical comment on practical computing, featuring the most important algorithms available at the time. It is not surprising that an early and visionary work would express opinions which would not hold up to future developments. He explained why Taylor series methods are not a good idea but today they have important roles in practical computation (see classic papers on Taylor series [1], [25]). He also did not think much of backward difference methods; these have since become the central component of generations of useful codes from DIFSUB to DASSL [3]. My own dealings with Peter were cordial and agreeable but I never got him to recognise that New Zealand is a different place from Australia and that this actually mattered to the inhabitants of these two countries separated by 2000km of ocean.

T. E. Hull

Tom Hull was a leading figure in numerical ordinary differential equations in Toronto and in Canada as a whole. His enduring impact rests on the DETEST and STIFF-DETEST projects [32], [18]. The idea was to provide standardized tests for the performance of differential equation solvers. Ultimately this led to mature test sets with wide international support based first at the CWI, Amsterdam, The Netherlands, and later at the University of Bari, Italy. Even when I first met him in 1965, Tom had already built a school of numerical analysis which has survived as an influential centre for numerical ordinary differential equations.

E. Hairer and G. Wanner

I first met these remarkable scientists in 1970 when, somewhat surprisingly, I got invited to a significant anniversary celebration at the University of Innsbruck. My series of lectures was basically an exposition of two papers [5], [8], the second of which was not yet published. Subsequently they wrote an important paper [28], in which the idea of B-series was developed. The relationship between this and my own paper was like the relationship between Taylor series and Taylor coefficients. This led to my own contribution becoming recognised, through their eyes, in a way that might otherwise have not been possible. In collaboration with S. P. Nørsett, whom I did not meet in person until much later, they introduced order stars [40]. Some of the achievements of this remarkable theoretical technique have already been referred to in Section 2.

G. Dahlquist

In my 1970 European visit I was privileged to meet Germund Dahlquist both in Stockholm and also at a summer school in France where we presented lectures on linear multistep methods (Germund) and Runge–Kutta methods (myself). I was already acquainted with the work in his famous thesis [13] and especially with the proof of the first Dahlquist barrier. The first proof I knew about was the one in Henrici's book [29] but there is a choice between sophistication at a high level or directness at a low level. Because the summer school organisers expected printed matter in addition to the verbal lectures, Germund, with the author's permission of course, distributed advance copies of the book by Bill Gear [22]. This formed a very nice introduction to many of the topics Germund was preparing to cover.

Without doubt, meeting and learning from, Germund Dahlquist, completed my personal introduction to the subject of numerical methods for differential equations.

References

1. Barton D., Willers I. M. and Zahar R. V. M., The automatic solution of systems of ordinary differential equations by the method of Taylor series, *Comput. J.*, **14** (1971), 243–248.

2. Bashforth F. and Adams J. C., *An Attempt to Test the Theories of Capillary Action by Comparing the Theoretical and Measured Forms of Drops of Fluid, with an Explanation of the Method of Integration Employed in Constructing the Tables which Give the Theoretical Forms of Such Drops.* Cambridge University Press, Cambridge (1883).

3. Brenan K. E., Campbell S. L. and Petzold L. R., *Numerical Solution of Initial-Value Problems in Differential-Algebraic Equations.* North-Holland, New York (1989).

4. Burrage K. and Butcher J. C., Non-linear stability of a general class of differential equation methods, *BIT*, **20** (1980), 185–203.

5. Butcher J. C., Coefficients for the study of Runge–Kutta integration processes, *J. Austral. Math. Soc.*, **3** (1963), 185–201.

6. Butcher J. C., A modified multistep method for the numerical integration of ordinary differential equations, *J. Assoc. Comput. Mach.*, **12** (1965), 124–135.

7. Butcher J. C., On the attainable order of Runge–Kutta methods, *Math. Comp.*, **19** (1965), 408–417.

8. Butcher J. C., An algebraic theory of integration methods, *Math. Comp.*, **26** (1972), 79–106.

9. Butcher J. C., A stability property of implicit Runge–Kutta methods, *BIT*, **15** (1975), 358–381.

10. Butcher J. C., Numerical methods for ordinary differential equations in the 20th century., *J. Comput. Appl. Math.*, **125** (2000), 1–29.

11. Butcher J C. and Wanner G., Runge-Kutta methods: some historical notes, *Appl. Numer. Math.*, **22** (1996), 113–151.

12. Curtiss C. F. and Hirschfelder J. O., Integration of stiff equations, *Proc. Nat. Acad. Sci. U.S.A.*, **38** (1952), 235–243.

13. Dahlquist G., Convergence and stability in the numerical integration of ordinary differential equations, *Math. Scand.*, **4** (1956), 33–53.

14. Dahlquist G., A special stability problem for linear multistep methods, *BIT*, **3** (1963), 27–43.

15. Dahlquist G., Error analysis for a class of methods for stiff non-linear initial value problems, *Lecture Notes in Math.*, **506** (1976), 60–72.

16. Dahlquist G., G-stability is equivalent to A-stability, *BIT*, **18** (1978), 384–401.
17. Deane J., *SILLIAC - Vacuum Tube Supercomputer.* Science Foundation for Physics (2006).
18. Enright W. H., Hull T. E. and Lindberg B., Comparing numerical methods for stiff systems of ODEs, *BIT*, **15** (1975), 10–48.
19. Euler L., De integratione aequationum differentialium per approximationem. In *Opera Omnia*, 1st series, Vol. 11, Institutiones Calculi Integralis, Teubner, Leipzig and Berlin, 424–434 (1913) .
20. Gear C. W., Hybrid methods for initial value problems in ordinary differential equations, *SIAM J. Numer. Anal.*, **2** (1965), 69–86.
21. Gear C. W., The numerical integration of ordinary differential equations, *Math. Comp.*, **21** (1967), 146–156.
22. Gear C. W., *Numerical Initial Value Problems in Ordinary Differential Equations.* Prentice Hall, Englewood Cliffs, NJ (1971).
23. Gear C. W., Algorithm 407, DIFSUB for solution of ordinary differential equations, *Comm. ACM*, **14** (1971), 185–190.
24. Gear C. W., Numerical solution of ordinary differential equations: is there anything left to do?, *SIAM Rev.*, **23** (1981), 10–24.
25. Gibbons A., A program for the automatic integration of differential equations using the method of Taylor series, *Comput. J.*, **3** (1960), 108–111.
26. Gill S., A process for the step-by-step integration of differential equations in an automatic computing machine, *Proc. Cambridge Philos. Soc.*, **47** (1951), 96–108.
27. Gragg W. B. and Stetter H. J., Generalized multistep predictor–corrector methods, *J. Assoc. Comput. Mach.*, **11** (1964), 188–209.
28. Hairer E. and Wanner G., On the Butcher group and general multi-value methods, *Computing*, **13** (1974), pp. 1–15.
29. Henrici P., *Discrete Variable Methods in Ordinary Differential Equations.* John Wiley & Sons Inc, New York (1962).
30. Heun K., Neue Methoden zur approximativen Integration der Differential- gleichungen einer unabhängigen Veränderlichen, *Z. Math. Phys.*, **45** (1900), 23–38.
31. Higham N. J., The accuracy of floating point summation, *SIAM J. Sci. Comput.*, **14** (1993), 783–799.
32. Hull T. E., Enright W. H., Fellen B. M. and Sedgwick A. E., Comparing numerical methods for ordinary differential equations, *SIAM J. Numer. Anal.*, **9** (1972), 603–637.
33. Kutta W., Beitrag zur näherungsweisen Integration totaler Differentialgleichungen, *Z. Math. Phys.*, **46** (1901), 435–453.
34. Merson R. H., An operational method for the study of integration processes. In *Proc. Symp. Data Processing*, Weapons Research Establishment, Salisbury, S. Aus- tralia (1957) .
35. Moulton F. R., *New Methods in Exterior Ballistics.* University of Chicago Press (1926).
36. Nordsieck A., On numerical integration of ordinary differential equations, *Math. Comp.*, **16** (1962), 22–49.
37. Obituary, A C Aitken, DSc, FRS, *Proc. Edinburgh Math. Soc.*, **16** (1968), 151–176.
38. Runge C., Über die numerische Auflösung von Differentialgleichungen, *Math. Ann.*, **46** (1895), 167–178.
39. Sadler D. H., Comrie, Leslie John. In Charles Couston *Dictionary of Scientific Biography*, Charles Scribner's Sons, NY (1970-80) .
40. Wanner G., Hairer E. and Nørsett S. P., Order stars and stability theorems, *BIT*, **18** (1978), 475–489.

Interview with Herbert Bishop Keller*

Hinke M. Osinga

Bristol Centre for Applied Nonlinear Mathematics
Department of Engineering Mathematics, University of Bristol,
Queen's Building, Bristol BS8 1TR, UK
h.m.osinga@bristol.ac.uk

Abstract. Herb Keller is well known in dynamical systems as the person who invented pseudo-arclength continuation. However, his work ranges as wide as scattering theory, fluid dynamics, and numerical analysis, and his publications span almost 60 years! At a workshop in Bristol, March 21-24, 2005, he shared some of his many stories with Hinke Osinga.

In March 2005 DSWeb Magazine had the splendid opportunity to do an interview with Herb Keller when he visited Bristol to attend the workshop Qualitative Numerical Analysis of High-dimensional Nonlinear Systems[1]. Herbert Bishop Keller, who was born in 1925 in Paterson, New Jersey, turns 80 this year ("Yes, in June... If I last that long"). Starting at the age of 20, he has been working in mathematics for almost 60 years and it would be impossible to convey the entire breadth and depth of his work in one article. This interview particularly focuses on his contributions in Dynamical Systems Theory.

Herbert B. Keller, March 24, 2005.
photographer Bernd Krauskopf.

Herb got a Bachelor in Electronic Engineering from Georgia Tech[2] in 1945. "This was a special program during the war years, where you took three semesters in one year for an otherwise four-year program. I did it in two years and eight months and so I was only 20 years old then." Herb got to Georgia Tech as part of the Naval Reserve Officer Training core program. He was trained as a fire control officer and he was on a battleship preparing for the invasion of Japan. There is a certain matter-of-fact-ness about it: "I got my diploma, and

---

* This article first appeared April 2005 in DSWeb Magazine, the quarterly newsletter of the Activity Group on Dynamical Systems of the Society for Industrial and Applied Mathematics; http://www.dynamicalsystems.org/ma/ma/display?item=108 They are acknowledged for their kind permission to reprint this interview.
[1] http://www.enm.bris.ac.uk/anm/workshop-c/
[2] Georgia Institute of Technology, Atlanta; http:www.gatech.edu

the next day I got married, and the next day I was off to war." From this point of view, one can understand the elation Herb felt when the bomb was dropped: it meant he could go home. "They had a point system: you could get discharged as soon as you had enough points." However, when Herb had collected the required number of points a few months later, he was told that he couldn't possibly leave! "I was their only qualified catapult officer, they couldn't lose me." As suggested by a fellow officer, Herb put a message on the notice board 'Catapult class, 8am Monday morning' and, indeed, four young men showed up. "They were all from Annapolis, and one of them was Jimmy Carter! After two weeks of training them, I could go home."

In 1946 Herb was back at Georgia Tech and trying to decide what to do. His two-year older brother Joe was by then a Faculty member at the Institute for Mathematics and Mechanics at New York University[3], which later was to become the Courant Institute of Mathematical Sciences. "I was interested in Engineering and Physics, so naturally I was good at Mathe-

Herb to Hinke: "I got my degree and the next day I got my commission as an Ensign in the Navy, and the next day I got my Georgia peach!" (March 24 2005)

photographer Bernd Krauskopf.

matics, but I had never thought about a career in Mathematics. Through my brother I began to believe that Mathematics might be interesting." Joe arranged for Herb to meet Courant. Obviously, an interview with Courant himself was a serious matter, and Herb decided to wear his Navy Whites. "I had worn that uniform only twice before, when I tried it on, and when I got married. It seemed appropriate to wear it again." It is unclear whether the uniform did it. Courant certainly was not much impressed with Herb's mathematical skills. "At one point he looked up and down the big windows in his office and asked: 'Do you know how to wash windows?' But he hired me anyway and, fortunately, he found something else for me to do!" So that was that. Herb went to New York and studied for his Master's, which he got in 1948.

---

[3] http://www.cims.nyu.edu

Joe and Herb Keller on la Tour Eiffel in 1948.

Herb wrote papers with his brother on scattering theory and the reflection and transmission of electromagnetic and sonic waves. Part-time he also taught at Sarah Lawrence College[4] in Bronxsville, New York. "There were 360 mostly wealthy and good-looking girls at this school, and I was the Math Department. That was a pretty nice job." As Herb was looking for a topic for his PhD, he realized that he should not work too closely with his brother. He decided to move away from scattering theory into computing. "I decided to become a Numerical Analysist." He wrote a thesis *On systems of linear ordinary differential equations with applications to ionospheric propagation* with Wilhelm Magnus as his advisor and received his PhD in 1954. From then on, he was a full-time member of the faculty at NYU.

NYU was an extremely active place. The Institute officially became the Courant Institute in 1960, with Niels Bohr as the dedication speaker. "Everyone from Europe who travelled to the US came through Courant. The institute was well known and I got to meet lots of people." In the late 1950s NYU bought the third *UNIVAC*. This is a 1000-word memory mercury delay line computer, the first commercially available high-speed computer. They founded the AEC Computing and Applied Mathematics Center, of which Herb was the associate director and Peter Lax was the director. "This was the beginning of Scientific Computing. I think I was very fortunate to be getting into the numerical business at the right time. Numerical Analysis really took off. The main problem was how to solve big systems on small computers. It is changing now, because there is so much good software. We didn't even know the term *software*!"

Herb and Joe Keller at Iguassue Falls in Brazil, 1985.

---

[4] http://www.slc.edu/

Research in the early 1960s always had a strong military flavor. "We had the cold war with Russia. We wanted to know how to protect people from bombs." Together with Bob Richtmeyer, Herb set up a research group working on nuclear reactions and the effects of atomic weapons. "We were interested in Mathematical Physics in much the same way as Los Alamos, but our stuff was not classified. We worked on the theory of nuclear reactors for naval power reactors. These were diffusion problems and transport problems. It was all done numerically."

At the same time, Herb was interested in fluid flow problems. "At Courant there was lots going on about fluid flow. You simply couldn't help being part of it. It really was the area where I started to work on computational things." He studied Von Karman swirling flow between rotating disks, Taylor-Couette flow, flows in channels and over airplane wings. "Oh yeah, and satellite re-entry problems. That is simply the flow around a sphere when entering the atmosphere at hypersonic speeds."

Developing a taste for dynamical system theory

At some point, Herb started working with Ed Reiss, who was interested in solid mechanics at Courant, on the buckling of spherical shells and rods and so on. "We treated it as an equilibrium phenomenon, while it was a bifurcation problem. But we didn't know that at the time. We realized it had dynamics, but didn't do much about it."

Herb Keller and Tony Chan (right) in China.

This was also the time when Herb moved from the Courant Institute to the department of Applied and Computational Mathematics[5] at Caltech. He had visited Caltech in 1965, but went back to NYU in 1966. "Courant came to see me and told me that it would be so much better for me to come back to NYU.

---

[5] http://www.acm.caltech.edu

I agreed with him. I mean, they had much more powerful machines than at Caltech. But when I came back to New York, I knew it was a mistake and I left for Caltech again in 1967 and have been there ever since."

At NYU, Herb had alternated teaching Numerical Analysis with Eugene Isaacson, who was quite a bit older than Herb. Wiley asked them whether they were interested in writing a book about it from the lecture notes. "We had a really good collaboration going, working on this book together. This continued while I was at Caltech, which is quite something, because we didn't have email at the time. I remember writing the preface of the book. I found it really difficult and wanted to get it over with. But when I sent it to Eugene, he responded that it was terrible and had all sorts of suggestions on how to change it. That's when I decided to work in my secret message. Do you know what an acronym is? You go and figure it out. And when you think you should give up, keep on reading! When Eugene saw it, he never complained again." [When Herb told me this, we were enjoying a sparkling Chardonnay at a nice restaurant, and I couldn't wait to get back to my office and read the preface. It's worth it.]

As Herb worked on the book with Eugene Isaacson, he got too carried away on two-point boundary value problems. "It was too much material and I decided to write a monograph on it separately. The techniques explained in that book really were the start of the continuation setting." At Caltech, in 1967, Herb learnt about Lyapunov-Schmidt methods, which made a big impression on him. It was then that he started doing serious bifurcation theory and path continuation. "The real turning point only came in 1976, when I was asked to speak at the *Advanced Seminar on Applications in Bifurcation Theory*, organized by Paul Rabinowitz and Mike Crandall. The army had a big center for research at the University of Wisconsin and they organized a series of meetings every year with a book for each. In my paper "Numerical Solution of Bifurcation and Nonlinear Eigenvalue Problems" for this book, I invented the term *pseudo-arclength continuation*. It is my most popular paper ever. Rabinowitz was very thankful, because the popularity of my paper meant that his book was very popular."

Stig Larsson, Simon Tavener, Herb Keller, Alastair Spence, Ridgway Scott, and Don Estep hiking in the Rocky Mountains outside of Fort Collins, Colorado, May 2001.

While research in Applied and Computational Mathematics at Caltech, which grew out of Aeronautics, was primarily focused on Fluid Dynamics, Herb brought in a strong numerical component. He built up a group of people working in Numerical Analysis, including Heinz-Otto Kreiss, Bengt Fornberg, Jens Lorenz, Tony Chan, Jim Varah, Eitan Tadmore and others. However, the core group at ACM was built by Gerald Whitham (a student of Lighthill's), and included Philip Saffman (who had worked with G.I. Taylor), Paco Lagerstrom (a topologist by training who worked on asymptotics in Aero), Julian Cole (also from Aero), Don Cohen (a student of Joseph Keller), and, of course, Herb himself. "In addition, I've had tons of excellent students and postdocs. You know, if you get garbage in, you get garbage out, but if you get good students in... It was a pleasure! Yes, in a sense I was lucky, being at such fantastic places and all that. I was influenced by the many good people around me. However, you cannot fake it in this business."

Retirement

In contrast to many other countries, there is no longer forced retirement in the US. In 1983, an old senator from Florida who himself was in the 90s, passed a law that one should not force people to retire just because of age. As a transition period, it was decided that in the following ten years, one must retire when reaching the age of 70. "I was lucky. My brother turned 70 just before the end of those ten years, but I didn't."

For many of the younger mathematicians, especially those who are still looking for a permanent job, it can be frustrating to see grey old men (mostly, isn't it!) sitting on precious academic posts. Herb ponders the possible benefits of forced retirement: "In fact, we did have a discussion at Caltech about how to get rid of the dead wood, but we decided that, at Caltech, we did not have to worry about it." Caltech has an appealing early retirement

Herb Keller with his grandson Milo at San Diego Zoo.

program, offering a two-year scholarly leave with full pay and no teaching (like a two-year sabbatical), at the end of which one should retire. "I think that good people do not hang on by the skin of their teeth. When they feel they do not pull their weight as compared to the people around them, they will pull out."

Herb retired from Caltech in 2001. He is now a research scientist at UC San Diego[6] and also kept his office at Caltech, which he also visits regularly.

---

[6] http://www.ucsd.edu

Jim Bunch, Herb Keller, and Randy Bank at UCSD, drinking beer.

"Aging has effects on your research. You need to do lots of calculations and I find it harder to do. It used to be fun, but it is not as much fun anymore and it certainly does not go as fast." Herb also got tired of teaching. "It is a big commitment and I certainly was not very good at it. I probably could have been more stimulating." Caltech annually awards the Feynman Prize for Excellence in Teaching. The winner is chosen by the students, and the prize consists of a cash award of $3,500 and an equivalent raise in the winner's salary. Herb has never won that prize ("Oh no, I didn't even come close!"), but he did ask advice on teaching from Harry Gray, the Beckman Professor of Chemistry, who is a very popular teacher. "He told me that you should never start lecturing straight away. The first five to ten minutes the students are not ready for you yet. Furthermore, they cannot concentrate for more than 15 minutes and it is impossible to have them absorb more than one new idea in a lecture. Well, I think he had quite a low opinion of the students, but he worked very hard at teaching professionally. I never thought about it. My concerns were how to best present the material and how to present the proofs." There is a second aspect: the rate at which the course content changes is often slow compared to the changes in technology. "You work so hard on getting the course material, you cannot keep up, certainly not if you ever want to get any research done."

"I must say that it feels great to be retired. We old guys thought we were irreplaceable, but we're not. We have been replaced and the young mathematicians are good. However, I do miss having students. That I miss most."

Addicted to cycling

Those who know Herb well know that he hardly ever goes anywhere without his bicycle. "I started cycling rather late, in early 1981 or so. After my divorce I wanted to keep more contact with my son. He was at UC Davis[7] and everyone bikes there, so..." After Herb had been invited to Oberwolfach, he decided to do a big bike tour in Germany and asked his son to join him. What should have been

---

[7] http://www.ucdavis.edu

a fantastic vacation ended dramatically when Herb could not make the turn on a downhill slope and ended up head first in a lumber yard. "It must have been terrible for my son. His first bike tour with his dad and then I had this dreadful accident." Herb spent one week in a hospital in Villingen-Schwenningen. Sebius Doedel [Sebius is enjoying the same Chardonnay in this restaurant] remembers the event quite well, because he paid Herb a visit at the hospital. Sebius tells us how he came to the hospital and asked the nurses where he could find Herb Keller. The nurses immediately knew who he meant, because Herb was their most honored guest: "Ah, der Herr Keller! Der ist mit einem Helikopter eingeflogen!!"

Despite this bad accident, Herb has kept cycling ever since. He has also had many other bicycling accidents, but none as remarkable as this first one. "It cured my eyesight! For 30 years I had glasses, even bi-focals, but I haven't worn glasses since."

Bristol, April 2005.

A personal perspective on the history of the numerical analysis of Fredholm integral equations of the second kind

Kendall Atkinson

The University of Iowa
atkinson@math.uiowa.edu

Abstract. This is a personal perspective on the development of numerical methods for solving Fredholm integral equations of the second kind, discussing work being done principally during the 1950s and 1960s. The principal types of numerical methods being studied were projection methods (Galerkin, collocation) and Nyström methods. During the 1950s and 1960s, functional analysis became the framework for the analysis of numerical methods for solving integral equations, and this influenced the questions being asked. This paper looks at the history of the analyses being done at that time.

1 Introduction

This memoir is about the history of the numerical analysis associated with solving the integral equation

$$\lambda x(s) - \int_a^b K(s,t)x(t)\,dt = y(s), \qquad a \le s \le b, \quad \lambda \ne 0. \tag{1.1}$$

At the time I was in graduate school in the early 1960's, researchers were interested principally in this one-dimensional case. It was for a kernel function K that was at least continuous; and generally it was assumed that $K(s,t)$ was several times continuously differentiable. This was the type of equation studied by Ivar Fredholm [26], and in his honor such equations are called *Fredholm integral equations of the second kind.*

Today we work with multi-dimensional Fredholm integral equations of the second kind in which the integral operator is completely continuous and the integration region is commonly a surface in $\mathbb{R}^3$; in addition, the kernel function K is often discontinuous. The Fredholm theory is still valid for such equations, and this theory is critical for the convergence and stability analysis of associated numerical methods. Throughout this paper, we assume the integral equation (1.1) is uniquely solvable for any given continuous function y.

The theory of Fredholm integral equations is quite old, and many such equations are associated with reformulations of elliptic boundary value problems as *boundary integral equations* (BIEs). More about BIEs later. Among the well-known names associated with the development of the theory of Fredholm integral equations are Ivar Fredholm and David Hilbert. For a survey of the origins of integral equations in applications, see Lonseth [44]. An interesting history of the origins of function spaces is given in Bernkopf [16]. He argues that much of the original motivation for creating function spaces arises from the study of integral equations (and secondarily, the calculus of variations).

A brief perusal of any mathematics library will yield many books on integral equations. An excellent introductory text on integral equations is that of Rainer Kress [41], and it also contains a good introduction to the numerical solution of integral equations.

2   A survey of numerical methods

There are only a few books on the numerical solution of integral equations as compared to the much larger number that have been published on the numerical solution of ordinary and partial differential equations. General books on the numerical solution of integral equations include, in historical order, Bückner [21], Baker [14], Delves and Mohamed [25], and Hackbusch [32]. Lonseth [43] gives a survey of numerical methods in the period preceding the widespread use of digital computers. For an interesting perspective on the theory, application, and numerical solution of nonlinear integral equations around the early 1960's, see the proceedings [1]. Important bibliographic references for both the application and the solution of integral equations are given in the very large subject and author listings of Noble [46].

Bückner's book was published in 1952, and it is representative of a pre-computer approach to the numerical analysis of integral equations. The book presents numerical methods for principally Fredholm integral equations of the second kind, with a shorter discussion of numerical methods for Volterra integral equations. The eigenvalue problem for integral operators is the principal focus of the book, with shorter treatments of some numerical methods for the inhomogeneous equation (1.1).

More specialized treatments of numerical methods for integral equations are given in the books Atkinson [9], [11], Brunner [20], Chatelin [23], Groetsch [29], Linz [42], Ivanov [35], and Wing [65]. Useful presentations of numerical methods are given in sections of [39, Chap. 14], [38, Chap. 2], [41, Chaps 10-17], and [12, Chaps 12, 13], along with sections of many general texts on integral equations. There are a number of edited proceedings, which we omit here. In the last 25 years, there has been a large amount of activity in numerical methods for solving boundary integral equation reformulations of partial differential equations. Introductions to this topic are given in [11], [32], and [41]. It is discussed briefly later in this paper.

Before discussing some of the history of the numerical analysis for (1.1), I give a brief survey of the general numerical methods for solving such integral equations. Most researchers subdivide the numerical methods into the following:

- Degenerate kernel approximation methods
- Projection methods
- Nyström methods (also called quadrature methods)

All of these methods have iterative variants, which I discuss briefly in §3.4. There are other numerical methods, but the above methods and their variants include the most popular general methods.

To expedite the presentation, I often use a functional analysis framework, even though such a presentation arose later in the history of these numerical methods. As an illustration, the integral equation (1.1) can be written abstractly as $(\lambda I - \mathcal{K}) x = y$ with $\mathcal{K}$ a compact integral operator on a Banach space $\mathcal{X}$, e.g. $C[a, b]$ or $L^2(a, b)$.

2.1   Degenerate kernel approximation methods

We say $K(s, t)$ is a *degenerate kernel function* if it has the form

$$K(s, t) = \sum_{j=1}^{n} \alpha_j(s) \beta_j(t).$$

In this case, the solution of (1.1) reduces to the solution of the linear system

$$\lambda c_i - \sum_{j=1}^{n} (\alpha_j, \beta_i) c_j = (y, \beta_i), \qquad i = 1, \ldots, n$$

and

$$x(s) = \frac{1}{\lambda} \left[y(s) + \sum_{j=1}^{n} c_j \alpha_j(s) \right]. \tag{2.1}$$

Most kernel functions $K(s, t)$ are not degenerate, and thus we seek to approximate them by degenerate kernels. We assume a sequence of degenerate kernels have been constructed, call them $K_n(s, t)$, for which

$$\max_{a \le s \le b} \int_a^b |K(s, t) - K_n(s, t)| \, dt \to 0 \quad \text{as} \quad n \to \infty. \tag{2.2}$$

Denote by x_n the result of solving the integral equation (1.1) with the approximate kernel K_n replacing K. For later reference, introduce the associated approximating integral operator

$$\mathcal{K}_n z(s) = \int_a^b K_n(s, t) z(t) \, dt, \quad a \le s \le b, \quad z \in \mathcal{X}.$$

Usually, $\mathcal{X}$ equals $C[a, b]$ or $L^2(a, b)$. Then x_n satisfies $(\lambda I - \mathcal{K}_n) x_n = y$; and if (1.1) is considered within the framework of the function space $C[a, b]$ with the uniform norm, then (2.2) is exactly the same as saying

$$\|\mathcal{K} - \mathcal{K}_n\| \to 0 \quad \text{as} \quad n \to \infty. \tag{2.3}$$

2.2   Projection methods

These methods approximate the solution x by choosing an approximation from a given finite dimensional linear subspace of functions, call it $\mathcal{Z}$. Given $z \in \mathcal{Z}$, introduce the residual

$$r = (\lambda I - \mathcal{K}) z - y.$$

We select a particular z, call it x^*, by making the residual r small in some sense. Let $\{\varphi_1, \ldots, \varphi_n\}$ denote a basis for $\mathcal{Z}$. Then we seek

$$x^*(s) = \sum_{j=1}^{n} c_j \varphi_j (s).$$

The residual becomes

$$r(s) = \sum_{j=1}^{n} c_j \{\lambda \varphi_j (s) - \mathcal{K} \varphi_j (s)\} - y(s).$$

- **Collocation method.** Select collocation node points $\{t_1, \ldots, t_n\} \in [a, b]$ and require

  $$r(t_i) = 0, \qquad i = 1, \ldots, n.$$

- **Galerkin method.** Set to zero the Fourier coefficients of r with respect to the basis $\{\varphi_1, \ldots, \varphi_n\}$,

  $$(r, \varphi_i) = 0, \qquad i = 1, \ldots, n.$$

  The basis $\{\varphi_i\}$ need not be orthogonal. The Galerkin method is also called the *method of moments*.

These are the principal projection methods, although there are others such as the minimization of the L^2 norm of r with respect to the elements in $\mathcal{Z}$.

With both collocation and Galerkin methods, it is possible to define a projection $\mathcal{P}$ with range $\mathcal{Z}$ and for which the numerical method takes the abstract form

$$(\lambda I - \mathcal{P}\mathcal{K}) x^* = \mathcal{P}y.$$

In practice we have a sequence of approximating subspaces $\mathcal{Z} = \mathcal{X}_n$, $n \geq 1$, and associated projections $\mathcal{P}_n$. Thus we have a sequence of approximating equations

$$(\lambda I - \mathcal{P}_n \mathcal{K}) x_n = \mathcal{P}_n y. \tag{2.4}$$

With Galerkin's method defined on a Hilbert space $\mathcal{X}$, $\mathcal{P}_n x$ is the orthogonal projection of x onto $\mathcal{X}_n$. For the collocation method, $\mathcal{P}_n x$ is the element of $\mathcal{X}_n$ which interpolates x at the node points $\{t_1, \ldots, t_n\}$.

We usually work with cases in which

$$\mathcal{P}_n z \to z \quad \text{as} \quad n \to \infty, \quad \text{for all} \quad z \in \mathcal{X} \tag{2.5}$$

although there are important cases where this is not satisfied. A weaker but adequate assumption is that the projections satisfy

$$\|\mathcal{K} - \mathcal{P}_n \mathcal{K}\| \to 0 \quad \text{as} \quad n \to \infty. \tag{2.6}$$

This also follows from (2.5) and the compactness of the operator $\mathcal{K}$. The space $\mathcal{X}$ is generally chosen to be $C[a, b]$ or $L^2(a, b)$, and we are solving (1.1) for the solution $x \in \mathcal{X}$. For details and examples, see [11, Chap. 3].

Projection methods are probably the most widely used class of methods for solving integral equations. For a presentation and summary of the most recent perspectives on projection methods for solving integral equations of the second kind, see [11, Chap. 3]. This also contains a discussion of 'discrete projection methods' in which integrals in the discretized linear system are replaced by numerical integrals.

2.3 Nyström methods

Approximate the integral operator in (1.1) using numerical integration. Consider a numerical integration scheme

$$\int_a^b f(t) \, dt \approx \sum_{j=1}^n w_j f(t_j)$$

which is convergent as $n \to \infty$ for all continuous functions $f \in C[a, b]$. Then introduce

$$\mathcal{K} z(s) \equiv \int_a^b K(s, t) z(t) \, dt$$
$$\approx \sum_{j=1}^n w_j K(s, t_j) z(t_j) \equiv \mathcal{K}_n z(s), \qquad a \le s \le b$$

for all $z \in C[a, b]$. We approximate the equation (1.1) by

$$(\lambda I - \mathcal{K}_n) x_n = y \tag{2.7}$$

or equivalently,

$$\lambda x_n(s) - \sum_{j=1}^n w_j K(s, t_j) x_n(t_j) = y(s), \qquad a \le s \le b. \tag{2.8}$$

This is usually solved by first collocating the equation at the integration node points and then solving the linear system

$$\lambda z_i - \sum_{j=1}^n w_j K(t_i, t_j) z_j = y(t_i), \qquad i = 1, \dots, n \tag{2.9}$$

in which $z_i \equiv x_n(t_i)$. Originally people would take this solution and then interpolate it in some way so as to extend it to the full interval $[a, b]$. However, it can be shown that the equation (2.8) furnishes a natural interpolation formula,

$$x_n(s) = \frac{1}{\lambda} \left[y(s) + \sum_{j=1}^{n} w_j K(s, t_j) z_j \right], \qquad a \le s \le b. \tag{2.10}$$

It turns out that this is a very good interpolation formula, as the resulting interpolated values have an accuracy that is comparable to that of the approximate solution $\{z_i\}$ at the integration node points. We solve (2.9) and may stop there with no interpolation; but for the theoretical analysis of the method, we use (2.8). This places the original equation (1.1) and the approximating equation (2.7) in the same function space, namely $C[a, b]$.

The interpolation formula (2.10) was noted by Nyström [48]. He was operating in an age of hand calculation, and therefore he wanted to minimize the need for such calculations. The great accuracy of (2.10) recommended itself, as then one could use a high accuracy integration rule with few nodes (e.g. Gaussian quadrature) while still having an accurate answer over the entire interval $[a, b]$. For that reason, and beginning in [4], I refer to the approximation (2.8) as the Nyström method. It has also been called the 'method of numerical integration' and the 'analogy method' (cf. [21, p. 105]). As a current example of an actual algorithmic use of Nyström interpolation, see [13].

With the rectangular numerical integration rule, (2.8) was used by Hilbert [33] in studying the symmetric eigenvalue problem for the integral operator $\mathcal{K}$. He used a limiting procedure to pass from the known properties of the symmetric eigenvalue problem for matrices to results for an integral operator $\mathcal{K}$ with a continuous and symmetric kernel function.

3   Error analysis and some history

The 1960's were a time of major change in numerical analysis, due in large part to the widespread introduction of digital computers. To obtain some sense of the contrast with today, consider my first numerical analysis course in 1961 and the text for it, Hildebrand [34]. This text was well-written, and it was fairly typical of numerical analysis textbooks of that time. The numerical methods were dominated by the need to do hand and desktop calculator computations. There was extensive material on finite differences and on methods that would make use of tables. By the mid-1960's there were several books in which digital computers were now the the main means of implementing methods, and this in turn led to a different type of numerical scheme. Pre-computer algorithms emphasized the use of tables and the use of the human mind to reduce the need for calculations. The use of computers led to the development of simpler methods in which the calculational power of the computer could profitably be brought to bear. For the numerical solution of integral equations such as (1.1), a major change was being able to solve much larger systems of linear equations.

This had been a major roadblock in the development of numerical methods for integral equations.

A major theme in theoretical numerical analysis in the 1950's and 1960's was the development of general frameworks for deriving and analyzing numerical methods, and such frameworks almost always used the language of functional analysis. This was true in many areas of numerical analysis and approximation theory, although I believe numerical linear algebra was less affected by this focus. Initially researchers were more interested in obtaining a better understanding of existing numerical methods than they were in creating new methods. The development of such abstract frameworks, led to the development of so-called 'optimal' numerical methods. Spline functions and finite element methods are both associated with this search for optimal methods. As the abstract frameworks solidified, they led to the development of new methods for new problems.

Especially important in the building of a more general and theoretical framework was the seminal paper of L. V. Kantorovich [37]. This paper was subsequently translated under the auspices of the U.S. National Bureau of Standards and deseminated fairly widely. The paper was quite long and consisted of several parts. A framework using functional analysis was given for the approximate solution of integral equations and other operator equations. Another part generalized the method of steepest descent to functionals over a Banach space. And yet another part developed a calculus for nonlinear operators on Banach spaces. This was followed by a generalization of Newton's method for solving nonlinear operator equations on Banach spaces. This paper was quite influential on me and many others. It took many years for the ideas in the paper to work their way through the research community. For easier access to the material in [37], see the book of Kantorovich and Akilov [39, Chaps. 14-18]. A related early book of importance for nonlinear integral equations is Krasnoselskii [40].

3.1   Degenerate kernel methods

The error analysis for degenerate kernel methods was well-understood without the need for a functional analysis framework, and carrying it over to function spaces was straightforward. Basically it is a consequence of the *geometric series theorem*. In particular, suppose an operator $\mathcal{A} : \mathcal{X} \to \mathcal{Y}$ is bounded, one-to-one and onto, with $\mathcal{X}$ and $\mathcal{Y}$ Banach spaces. Suppose $\mathcal{B} : \mathcal{X} \to \mathcal{Y}$ is bounded, and further assume that

$$\|\mathcal{A} - \mathcal{B}\| < 1/\left\|\mathcal{A}^{-1}\right\|. \qquad (3.1)$$

Then $\mathcal{B}$ is also one-to-one and onto, and its inverse $\mathcal{B}^{-1}$ is bounded. Moreover, $\left\|\mathcal{A}^{-1} - \mathcal{B}^{-1}\right\| = O\left(\|\mathcal{A} - \mathcal{B}\|\right)$. In the case of degenerate kernel methods with $\mathcal{A} = \lambda I - \mathcal{K}$ and $\mathcal{B} = \lambda I - \mathcal{K}_n$, and working within the context of $C[a, b]$, the bound (2.3) gives us a bound for $\|\mathcal{A} - \mathcal{B}\|$. More precisely, if

$$\|\mathcal{K} - \mathcal{K}_n\| < \frac{1}{\left\|(\lambda I - \mathcal{K})^{-1}\right\|}$$

then $(\lambda I - \mathcal{K}_n)^{-1}$ exists and it can be bounded uniformly for all sufficiently large n. Letting $(\lambda I - \mathcal{K}) x = y$ and $(\lambda I - \mathcal{K}_n) x_n = y$,

$$\|x - x_n\| \leq \left\| (\lambda I - \mathcal{K}_n)^{-1} \right\| \|\mathcal{K} - \mathcal{K}_n\| \|x\|.$$

This leads to a straightforward error analysis for the degenerate kernel method when considered within the function space $C[a, b]$ using the uniform norm.

This basic analysis is given in many textbooks when developing the theory of integral equations, although historically if was often without the functional analysis framework; and it was often also used to develop some of the theory of the eigenvalue problem for integral operators. Since the degenerate kernel method was used both as a theoretical tool in developing the theory of integral equations and as a numerical method, it is difficult to give attributions to the development of the numerical method. In talking about the numerical method, much time has been spent on developing various means of approximating general kernel functions $K(s,t)$ with a sequence of degenerate kernels $K_n(s,t)$, and this continues to the present day. For illustrative examples of the degenerate kernel method, see [11, Chap. 2].

3.2 Projection methods

The general framework for projection methods and other approximation methods that was given by Kantorovich [37] was too complicated when considering only projection methods. Later work simplified his framework a great deal, and new perspectives continued to be given well into the 1980's.

The general error analysis for projection methods uses the assumption (2.6) that $\|\mathcal{K} - \mathcal{P}_n\mathcal{K}\| \to 0$ as $n \to \infty$. With the assumption that $(\lambda I - \mathcal{K})^{-1}$ exists, we write

$$\lambda I - \mathcal{P}_n\mathcal{K} = (\lambda I - \mathcal{K}) + (\mathcal{K} - \mathcal{P}_n\mathcal{K})$$
$$= (\lambda I - \mathcal{K}) \left[I - (\lambda I - \mathcal{K})^{-1} (\mathcal{K} - \mathcal{P}_n\mathcal{K}) \right]. \tag{3.2}$$

With the assumption (2.6), we have that

$$\left\| (\lambda I - \mathcal{K})^{-1} \right\| \|\mathcal{K} - \mathcal{P}_n\mathcal{K}\| < 1$$

for all sufficiently large n. It then follows from the geometric series theorem that $\left[I - (\lambda I - \mathcal{K})^{-1} (\mathcal{K} - \mathcal{P}_n\mathcal{K}) \right]^{-1}$ exists and is bounded, and therefore the same is true for $(\lambda I - \mathcal{P}_n\mathcal{K})^{-1}$. For the error, let $(\lambda I - \mathcal{K}) x = y$ and $(\lambda I - \mathcal{P}_n\mathcal{K}) x_n = \mathcal{P}_n y$. Then

$$x - x_n = \lambda (\lambda I - \mathcal{P}_n\mathcal{K})^{-1} (x - \mathcal{P}_n x). \tag{3.3}$$

This implies

$$\frac{|\lambda|}{\|\lambda - \mathcal{P}_n\mathcal{K}\|} \|x - \mathcal{P}_n x\| \leq \|x - x_n\| \leq |\lambda| \left\| (\lambda - \mathcal{P}_n\mathcal{K})^{-1} \right\| \|x - \mathcal{P}_n x\|. \tag{3.4}$$

We have convergence if and only if $P_n x \to x$ as $n \to \infty$. The speed of convergence of x_n to x is precisely the same as that of $P_n x$ to x.

A number of researchers have contributed to this theory and to extensions not discussed here. There are many papers on collocation methods for solving Fredholm integral equations of the second kind. For a general framework within a functional analysis framework, I cite in particular those of Phillips [50] and Prenter [51].

3.2.1 Kantorovich and Krylov regularization

An early and interesting variant on projection methods was given by Kantorovich and Krylov [38, p. 150]. As with projection methods, suppose a family of approximating functions $\mathcal{Z}$ is given with basis $\{\varphi_1, \ldots, \varphi_n\}$. Assume a solution for (1.1) of the form

$$x^*(s) = \frac{1}{\lambda}\left[y(s) + \sum_{j=1}^{n} c_j \varphi_j(s) \right]. \tag{3.5}$$

This was motivated, perhaps, by the solution (2.1) for a degenerate kernel integral equation. In effect we are seeking an approximation of the integral operator term $\mathcal{K}x$ in (1.1),

The authors looked at the residual r for such an approximating solution and then minimized it in the same manner as with Galerkin's method (although collocation methods can be used equally well). Introduce

$$z^*(s) = \sum_{j=1}^{n} c_j \varphi_j(s)$$

substitute it into the formula for $r = (\lambda I - \mathcal{K}) x^* - y$, and then minimize r with either a Galerkin or collocation method. Then z^* satisfies $(\lambda I - \mathcal{P}\mathcal{K}) z^* = \mathcal{P}\mathcal{K}y$ and x^* satisfies $(\lambda I - \mathcal{P}\mathcal{K}) x^* = y$. Because $\mathcal{P}$ has a finite-dimensional range, and because we always assume $\mathcal{P}$ is bounded, the combined operator $\mathcal{P}\mathcal{K}$ can be shown to be an integral operator with a degenerate kernel function. Thus the assumption (3.5) amounts to the approximation of (1.1) by a degenerate kernel integral equation.

Another way of picturing this method is to consider (1.1) in the form

$$x = \frac{1}{\lambda}(y + z), \qquad z = \mathcal{K}x. \tag{3.6}$$

The function $\mathcal{K}x$ is often better behaved than the original solution x, and this is particularly true if x is badly behaved (e.g. lacking differentiability at points in $[a, b]$). The function z satisfies the equation

$$(\lambda I - \mathcal{K}) z = \mathcal{K}y. \tag{3.7}$$

Applying a projection method to this equation and then using (3.6) leads to the method of Kantorovich and Krylov. The use of the formulation (3.6)-(3.7) is often referred to as the Kantorovich and Krylov method of regularizing the integral equation $(\lambda I - \mathcal{K}) x = y$.

3.2.2   The iterated projection solution Associated with the projection method solution x_n is the *iterated projection solution*. Given the projection method solution x_n, define

$$\widetilde{x}_n = \frac{1}{\lambda}\left[y + \mathcal{K}x_n\right]. \tag{3.8}$$

Although such iterations are found in the literature in many places, Ian Sloan [54] first recognized the importance of doing one such iteration; and in his honor $\widetilde{x}_n$ is often called the *Sloan iterate*.

The solution $\widetilde{x}_n$ satisfies the equation

$$(\lambda I - \mathcal{K}\mathcal{P}_n)\,\widetilde{x}_n = y \tag{3.9}$$

and $\mathcal{P}_n\widetilde{x}_n = x_n$. It can be shown that $(\lambda I - \mathcal{K}\mathcal{P}_n)^{-1}$ exists if and only if $(\lambda I - \mathcal{P}_n\mathcal{K})^{-1}$ exists; cf. [11, §3.4].

In the case of the Galerkin method over a Hilbert space $\mathcal{X}$, Sloan showed that the iterated solution $\widetilde{x}_n$ converges to x more rapidly than does the original Galerkin solution x, provided $\mathcal{P}_n$ is pointwise convergent to the identity I on $\mathcal{X}$ (as in (2.5)). Begin with the identity

$$x - \widetilde{x}_n = (\lambda - \mathcal{K}\mathcal{P}_n)^{-1}\mathcal{K}(I - \mathcal{P}_n)x. \tag{3.10}$$

Note that $I - \mathcal{P}_n$ is a projection, and therefore

$$\mathcal{K}(I - \mathcal{P}_n)x = \mathcal{K}(I - \mathcal{P}_n)(I - \mathcal{P}_n)x$$
$$\|\mathcal{K}(I - \mathcal{P}_n)x\| \le \|\mathcal{K}(I - \mathcal{P}_n)\|\,\|(I - \mathcal{P}_n)x\|.$$

With Galerkin's method, $I - \mathcal{P}_n$ is an orthogonal projection and is self-adjoint. Also, the norm of an operator on a Hilbert space equals that of its adjoint. Therefore,

$$\|\mathcal{K}(I - \mathcal{P}_n)\| = \left\|[\mathcal{K}(I - \mathcal{P}_n)]^*\right\| = \|(I - \mathcal{P}_n)^*\mathcal{K}^*\|$$
$$= \|(I - \mathcal{P}_n)\mathcal{K}^*\|.$$

If the operator $\mathcal{K}$ is compact on $\mathcal{X}$, then so is $\mathcal{K}^*$; and when combined with (2.5), we have $\|(I - \mathcal{P}_n)\mathcal{K}^*\| \to 0$. Completing the error bound,

$$\|x - \widetilde{x}_n\| \le \left\|(\lambda - \mathcal{K}\mathcal{P}_n)^{-1}\right\|\,\|\mathcal{K}(I - \mathcal{P}_n)x\|$$
$$\le c\,\|(I - \mathcal{P}_n)\mathcal{K}^*\|\,\|(I - \mathcal{P}_n)x\|.$$

When compared with the earlier result (3.4), this shows the earlier assertion that the iterated solution $\widetilde{x}_n$ converges to x more rapidly than does the original Galerkin solution x.

For collocation methods, we do not have $\|\mathcal{K}(I - \mathcal{P}_n)\| \to 0$. Nonetheless, the Sloan iterated solution $\widetilde{x}_n$ is still useful. From the property $\mathcal{P}_n\widetilde{x}_n = x_n$, we know that x_n and $\widetilde{x}_n$ agree at the node points $\{t_1, \ldots, t_n\}$. Thus an error bound for $\|x - \widetilde{x}_n\|_\infty$ is also a bound on the error

$$E_n = \max_{1 \le i \le n} |x(t_i) - x_n(t_i)|.$$

To bound E_n, we can use the formula (3.10) and analyze $\|\mathcal{K}(I - \mathcal{P}_n)x\|_\infty$. Using this, Graeme Chandler in his thesis [22] showed that astute choices of interpolation nodes (e.g. Gauss-Legendre zeroes) led to $E_n \to 0$ at a rate that was faster than the speed with which $\|x - x_n\|_\infty \to 0$. The collocation points $\{t_1, \ldots, t_n\}$ are said to be points of superconvergence with respect to the solution x_n over $[a, b]$.

3.3 Nyström methods

A central feature of the error analysis for degenerate kernel and projection methods is the justifiable assumption that $\|\mathcal{K} - \mathcal{K}_n\| \to 0$ as $n \to \infty$, where $\mathcal{K}_n$ denotes the associated approximation of the integral operator $\mathcal{K}$. With degenerate kernel methods, $\mathcal{K}_n$ is a degenerate kernel integral operator; and for projection methods, $\mathcal{K}_n = \mathcal{P}_n \mathcal{K}$. As discussed above, this leads to a straightforward error analysis based on the geometric series theorem.

In contrast, quadrature-based discretizations satisfy the relation

$$\|\mathcal{K} - \mathcal{K}_n\| \geq \|\mathcal{K}\|, \qquad n \geq 1.$$

As a consequence, the convergence analysis for the Nyström method must be something different than that used for degenerate kernel methods and projection methods.

The first convergence analysis for the Nyström method, to this author's knowledge, was given by Kantorovich and Krylov [38, p. 103]. Their analysis is complicated, but it is complete and is equivalent to the bounds of some later authors. Its significance appears to have been overlooked by later researchers. The 1948 paper of Kantorovich [37] contains a general schema for analyzing discretizations of operator equations, and using it he gives another convergence analysis for the solution at the node points. Yet another early analysis is given by Bückner [21] using arguments related to those for degenerate kernel methods and using piecewise constant interpolation to extend the nodal solution to the full interval.

For an approach that leads to the way in which the Nyström method is currently analysed, begin by defining

$$E_n(s, t) = \int_a^b K(s, v)K(v, t)dv - \sum_{j=1}^n w_j K(s, t_j)K(t_j, t). \tag{3.11}$$

With continuous kernel functions K and standard quadrature schemes that are convergent on $C[a, b]$, Mysovskih [45] showed that

$$E_n(s, t) \to 0 \quad \text{as} \quad n \to \infty \tag{3.12}$$

uniformly in (s, t). He used this to give a more transparent convergence analysis for Nyström's method. The convergence result (3.12) shows, implicitly, that in the context of $C[a, b]$, we have that

$$\|(\mathcal{K} - \mathcal{K}_n)\mathcal{K}\|, \ \|(\mathcal{K} - \mathcal{K}_n)\mathcal{K}_n\| \to 0 \quad \text{as} \quad n \to \infty. \tag{3.13}$$

This follows easily from the formulas

$$\|(\mathcal{K} - \mathcal{K}_n)\,\mathcal{K}\| = \max_{a \leq s \leq b} \int_a^b |E_n\,(s,t)|\,dt$$

$$\|(\mathcal{K} - \mathcal{K}_n)\,\mathcal{K}_n\| = \max_{a \leq s \leq b} \sum_{j=1}^n |w_j E_n\,(s,t_j)|. \tag{3.14}$$

Anselone and Moore [3] were interested in freeing the error analysis from the specific form of the integral equation (1.1) and its approximation (2.8). They found that such an argument using $E_n\,(s,t)$ in (3.14) could be avoided within an operator theoretic framework that was based on the following three assumptions.

A1. $\mathcal{K}$ and $\mathcal{K}_n$, $n \geq 1$, are bounded linear operators from a Banach space $\mathcal{X}$ into itself.
A2. $\mathcal{K}_n x \to \mathcal{K}x$ as $n \to \infty$ for all $x \in \mathcal{X}$.
A3. $\{\mathcal{K}_n x : n \geq 1$ and $\|x\| \leq 1\}$ has compact closure in $\mathcal{X}$.

From these hypotheses, it follows that $\|(\mathcal{K} - \mathcal{K}_n)\,\mathcal{K}\|$ and $\|(\mathcal{K} - \mathcal{K}_n)\,\mathcal{K}_n\|$ converge to zero as $n \to \infty$. To prove this, begin by letting B denote the set in **A3**; its closure $\overline{B}$ is compact. Then

$$\|(\mathcal{K} - \mathcal{K}_n)\,\mathcal{K}_n\| = \sup_{\|x\| \leq 1} \|(\mathcal{K} - \mathcal{K}_n)\,\mathcal{K}_n x\|$$

$$\leq \sup_{z \in \overline{B}} \|\,(\mathcal{K} - \mathcal{K}_n)\,z\|.$$

In addition, **A3** implies the family $\{\mathcal{K}_n\}$ is uniformly bounded; and thus it is an equicontinuous family on any bounded subset of $\mathcal{X}$. It is straightforward to show that pointwise convergence of a sequence of functions on a compact set is uniform. Combining these results leads to the convergence $\|(\mathcal{K} - \mathcal{K}_n)\,\mathcal{K}_n\| \to 0$.

An approximating family $\{\mathcal{K}_n\}$ satisfying **A1-A3** is said to be 'pointwise convergent and collectively compact'. This framework turns out to be quite important as there are important extensions of the standard Nyström approximation (2.8) for which one cannot show directly, as in (3.14), that $\|(\mathcal{K} - \mathcal{K}_n)\,\mathcal{K}\|$ and $\|(\mathcal{K} - \mathcal{K}_n)\,\mathcal{K}_n\|$ converge to zero. Product quadrature methods for treating singular kernel functions are examples. In addition, the family of approximating operators $\{\mathcal{K}_n\}$ for both degenerate kernel and projection methods satisfy **A3**.

In the error analysis, the earlier argument (3.2) is replaced by the following:

$$\frac{1}{\lambda}\left\{I + (\lambda I - \mathcal{K})^{-1}\mathcal{K}_n\right\}(\lambda I - \mathcal{K}_n) = I + \frac{1}{\lambda}(\lambda I - \mathcal{K})^{-1}(\mathcal{K} - \mathcal{K}_n)\mathcal{K}_n. \tag{3.15}$$

This identity originates from the following.

$$(\lambda I - \mathcal{S})^{-1} = \frac{1}{\lambda}\left[I + (\lambda I - \mathcal{S})^{-1}\,\mathcal{S}\right]$$

$$\approx \frac{1}{\lambda}\left[I + (\lambda I - \mathcal{T})^{-1}\,\mathcal{S}\right]$$

$$\frac{1}{\lambda} \left[I + (\lambda I - \mathcal{T})^{-1} \mathcal{S} \right] (\lambda I - \mathcal{S}) = I + \frac{1}{\lambda} (\lambda I - \mathcal{T})^{-1} (\mathcal{T} - \mathcal{S}) \mathcal{S}.$$

We assume $(\lambda I - \mathcal{K})^{-1}$ exists. From (3.15) and using $\|(\mathcal{K} - \mathcal{K}_n) \mathcal{K}_n\| \to 0$, we can show that $(\lambda I - \mathcal{K}_n)^{-1}$ exists and is uniformly bounded for all sufficiently large n. We can solve (3.15) for $(\lambda I - \mathcal{K}_n)^{-1}$ to obtain bounds on it that are uniform for sufficiently large n. Letting $(\lambda I - \mathcal{K}) x = y$ and $(\lambda I - \mathcal{K}_n) x_n = y$, we have

$$x - x_n = (\lambda I - \mathcal{K}_n)^{-1} (\mathcal{K}x - \mathcal{K}_n x). \tag{3.16}$$

This shows that the speed of convergence of x_n to x is at least as rapid as the speed of convergence of $\mathcal{K}_n x$ to $\mathcal{K}x$. A more complete discussion of collectively compact operator approximation theory is given in [2].

Other researchers developed related ideas, and the best known are probably those of Stummel [58] (involving the concept of *discrete compactness*) and Vainikko [60] (involving the concept of *compact approximation*). Their frameworks are more general in that the approximating equations can be defined on separate Banach spaces $\mathcal{X}_n$, $n \geq 1$. Another approach to understanding the Nyström method was given by Noble [47]. He developed an alternative framework using the language of prolongation and restriction operators, in some ways reminiscent of the original work of Kantorovich [37], but simpler and with new insights.

Using the abstract framework of collectively compact operators, a number of extensions of the Nyström method have been analyzed. We discuss two of them.

3.3.1 Product integration Consider singular kernel functions such as $K(s,t) = \log |s - t|$ or $|s - t|^{-\alpha}$, $\alpha < 1$. These define compact integral operators on $C[a, b]$; but an approximation of the associated integral operator based on standard numerical integration is a poor idea. To introduce the main idea of product integration, consider the particular kernel function

$$K(s,t) = L(s,t) \log |s - t|$$

with $L(s,t)$ a well-behaved kernel function.

As a particular case of product integration, let

$$a = t_0 < t_1 < \cdots < t_n = b.$$

Let $[z(t)]_n$ denote the piecewise linear interpolant to $z(t)$ with node points $\{t_0, \ldots, t_n\}$. Define the approximation of $\mathcal{K}x$ by

$$\mathcal{K}x(s) \approx \int_a^b [L(s,t)x(t)]_n \log |s - t| \, dt \equiv \mathcal{K}_n x(s).$$

It is straightforward to show that

$$\mathcal{K}_n x(s) = \sum_{j=0}^n w_j(s) L(s, t_j) x(t_j). \tag{3.17}$$

This is often called the product trapezoidal approximation.

The approximating equation $(\lambda I - \mathcal{K}_n)\, x_n = y$ can be dealt with in exactly the same manner as in (2.7)-(2.10) for the original Nyström method. These ideas for product integration were introduced in [4], [6], motivated in part by Young [66].

For the error analysis, the family $\{\mathcal{K}_n\}$ can be shown to be a pointwise convergent and collectively compact family. The error analysis reduces to that already done for such families, with bounds on the speed of convergence obtainable from (3.16). It should be noted that it was not possible to show directly for (3.17) the required convergence in (3.13).

Much research has been done on product integration methods. De Hoog and Weiss [24] give asymptotic error estimates when $x(t)$ is a smooth function. Among the more important results are those showing that the solution x of such equations $(\lambda I - \mathcal{K})\, x = y$ are usually poorly behaved around the endpoints of $[a, b]$; cf. Graham [27], Richter [52], and Schneider [53]. The latter paper [53] also discusses how to grade the mesh $\{t_i\}$ so as to compensate for the bad behaviour in the solution around the endpoints. For a general discussion of product integration, see [11, §4.2].

3.3.2   The eigenvalue problem Consider finding the eigenvalues λ and eigenfunctions $x \neq 0$ for the compact integral operator $\mathcal{K}$,

$$\int_a^b K(s,t) x(t)\, dt = \lambda x(s), \qquad a \leq s \leq b, \quad \lambda \neq 0. \tag{3.18}$$

This is a very old problem, and there is a large research literature on both it and its numerical solution. For a bibliography and some discussion of the early literature, see Bückner [21]. Among the papers in the research literature on the numerical solution of the problem, I particularly note [5], [8], Brakhage [18], Bramble and Osborn [19], Vainikko [59], and Wielandt [64]. The book of Chatelin [23, Chap. 4] gives a general presentation which includes a number of these results, and it contains an up-to-date bibliography of the field.

Let $\lambda_0 \neq 0$ be an eigenvalue of $\mathcal{K}$. Let $\varepsilon > 0$ be chosen so that the set $\mathcal{F} \equiv \{\lambda : |\lambda - \lambda_0| \leq \varepsilon\}$ contains no other eigenvalues of $\mathcal{K}$ and also does not contain 0. Let $\{\mathcal{K}_n\}$ be a collectively compact and pointwise convergent family of approximations to $\mathcal{K}$ on a Banach space $\mathcal{X}$ Let

$$\sigma_n = \left\{ \lambda_1^{(n)}, \ldots, \lambda_{r_n}^{(n)} \right\}$$

denote the eigenvalues of $\mathcal{K}_n$ that are located in $\mathcal{F}$. It can be shown that for n sufficiently large, σ_n is contained in the interior of $\mathcal{F}$.

Define

$$E\left(\lambda_0\right) = \frac{1}{2\pi i} \int_{|\mu - \lambda_0| = e} (\mu I - \mathcal{K})^{-1}\, d\mu.$$

$E(\lambda_0)$ is the *spectral projection* associated with λ_0. $E(\lambda_0)\mathcal{X}$ is the finite-dimensional subspace of simple and generalized eigenvectors associated with the eigenvalue λ_0 for $\mathcal{K}$,

$$E(\lambda_0)\mathcal{X} = \text{Null}\left((\lambda_0 I - \mathcal{K})^{\nu(\lambda_0)}\right)$$

with $\nu(\lambda_0)$ the index of the eigenvalue λ_0.
 Define

$$E_n(\sigma_n) = \frac{1}{2\pi i}\int_{|\mu-\lambda_0|=\varepsilon}(\mu I - \mathcal{K}_n)^{-1}\,d\mu.$$

$E_n(\sigma_n)\mathcal{X}$ is the direct sum of the subspaces of the simple and generalized eigenvectors associated with the eigenvalues of $\mathcal{K}_n$ contained in the approximating set σ_n,

$$E_n(\sigma_n)\mathcal{X} = \text{Null}\left(\left(\lambda_1^{(n)}I - \mathcal{K}\right)^{\nu\left(\lambda_1^{(n)}\right)}\right)\oplus\cdots\oplus\text{Null}\left(\left(\lambda_r^{(n)}I - \mathcal{K}\right)^{\nu\left(\lambda_r^{(n)}\right)}\right).$$

It is shown in [5] that the approximating eigenvalues in σ_n converge to λ_0,

$$\max_{\lambda\in\sigma_n}|\lambda - \lambda_0| \to 0 \quad\text{as}\quad n\to\infty.$$

Also, for every simple and generalized eigenvector $\varphi\in E(\lambda_0)\mathcal{X}$,

$$E_n(\sigma_n)\varphi\to\varphi \quad\text{as}\quad n\to\infty.$$

The element $E_n(\sigma_n)\varphi$ is a sum of simple and generalized eigenvectors associated with approximating eigenvalues in σ_n. Error bounds are also given in [5], [8]; and Bramble and Osborn [19] give both bounds and a beautifully simple way to improve the convergence in the case of an eigenvalue with multiplicity greater than 1. These results also apply to degenerate kernel methods and projection methods since the associated approximations $\mathcal{K}_n$ can be shown to be collectively compact and pointwise convergent.

 Related and independent error analysis results are given by Vainikko [59], and they are for a more general discretization framework than that of Anselone and Moore.

3.4   Iterative variants

There are iterative variants of all of the numerical methods discussed above. The linear systems for all of these numerical methods result in dense linear systems, say of order n, and then the cost of solution is $O(n^3)$. In addition, with both degenerate kernel methods and projection methods, the elements of the coefficient matrix are integrals which are usually evaluated numerically. With the collocation method these coefficients are single integrals, and with Galerkin's method, they are double integrals. The cost of evaluating the coefficient matrix is generally $O\left(n^2\right)$, although the constant of proportionality may be quite large. Evaluating the coefficient matrix for a Nyström method is also $O\left(n^2\right)$, but now each coefficient is only a single evaluation of the kernel function.

Most standard iteration methods for solving linear systems of order n, including Krylov subspace methods, lead to a cost of $O\left(n^2\right)$, which is consistent with the cost of setting up the coefficient matrix. Two-grid methods were introduced by Brakhage [17] and then developed much more extensively in [7], [9] for both linear and nonlinear problems. These methods also have a cost of $O\left(n^2\right)$. In [30] Hackbusch developed a fast multigrid iteration method with a cost of $O\left(n\right)$ for solving these linear systems; but the cost of setting up the linear system is still $O\left(n^2\right)$. For a much more extensive look at iterative variants of the methods of this paper, see [11, Chap. 6].

4   Boundary integral equation methods

A major use of integral equations has been to reformulate problems for partial differential equations (PDEs), and historically this dates back over 150 years. For example, the classical Neumann series (circa 1873) for solving integral equations refers to work of Carl Neumann in which he was considering a   problem in potential theory.   For a more complete look at this historical background, see Bateman [15] and Lonseth [44].

Along with the growth from the 1950's onward of finite element methods for solving elliptic PDEs, there was also interest in developing 'boundary element methods' (BEM) for solving 'boundary integral equation' (BIE) reformulations of elliptic PDEs. These integral equation reformulations reduce by 1 the dimensionality of the boundary value problem; and sometimes the solution of interest is needed only on the boundary of the region on which the original PDE is defined. The engineering literature on BEM is enormous, and there are several annual meetings on various aspects of the topic. In the community of researchers devoted to the numerical solution of Fredholm integral equations, the numerical solution of BIE has been a major focus from the late 1970's to the present day.

There are a number of ways to approach the development of BIE reformulations and their numerical analysis, and my perspective is biased by my own work in the area. Among introductions, I refer the reader to the books of [11, Chaps. 7-9], [32], [36], [41], [49], and [63]; for planar BIE problems, see the extensive survey article of Sloan [55]. A survey of the numerical solution of BIE for Laplace's equation is given in [10].

There are many outstanding research papers, and I can only refer to a few of them here; see [11] for a much more extensive bibliography. In his 1968 paper [62], Wendland laid the foundation for collocation methods for solving BIE reformulations in $\mathbb{R}^3$. In [31], Hackbusch and Nowak gave a fast way to set up and solve discretizations of BIE, with a total cost of $O\left(n\log^d n\right)$, d a small integer. An alternative fast method of solution, the fast multipole method, is given by Greengard and Rokhlin [28].

A true extension of the finite element method from PDE to BIE, retaining its variational framework, was introduced by Stephan and Wendland in [57]. This included BIE that were integral equations of the first kind ($\lambda = 0$ in (1.1)) as well as of the second kind. This framework opened a very fruitful approach to solving BIE, and it is still a very active area of research. An extended discussion of this finite element method is given in [63], and there is a large research literature on it.

Additional comments. Although a number of papers are given in the following bibliography, a much more complete list is given in [11], and additional discussions of the literature are given at the conclusions of the various chapters therein.

Acknowledgements. I thank the organizers of this conference, *The Birth of Numerical Analysis*, for their efforts in making it such a success and for giving me the chance to participate in it.

References

1. P. Anselone, ed., *Nonlinear Integral Equations*, Univ. of Wisconsin Press, 1964.
2. P. Anselone, *Collectively Compact Operator Approximation Theory and Applications to Integral Equations*, Prentice-Hall, 1971.
3. P. Anselone and R. Moore, Approximate solution of integral and operator equations, *J. Math. Anal. Appl.* **9** (1964), 268–277.
4. K. Atkinson, *Extensions of the Nyström Method for the Numerical Solution of Integral Equations of the Second Kind*, Ph.D. dissertation, Univ. of Wisconsin, Madison, 1966.
5. K. Atkinson, The numerical solution of the eigenvalue problem for compact integral operators, *Trans. Amer. Math. Soc.* **129** (1967), 458–465.
6. K. Atkinson, The numerical solution of Fredholm integral equations of the second kind, *SIAM J. Num. Anal.* **4** (1967), 337–348.
7. K. Atkinson, Iterative variants of the Nyström method for the numerical solution of integral equations, *Numer. Math.* **22** (1973), 17–31.
8. K. Atkinson, Convergence rates for approximate eigenvalues of compact integral operators, *SIAM J. Num. Anal.* **12** (1975), 213–222.
9. K. Atkinson, *A Survey of Numerical Methods for the Solution of Fredholm Integral Equations of the Second Kind*, SIAM, Philadelphia, 1976.
10. K. Atkinson, The numerical solution of boundary integral equations, in *The State of the Art in Numerical Analysis*, ed. by I. Duff and G. Watson, Clarendon Press, Oxford, 1997, 223–259.
11. K. Atkinson, *The Numerical Solution of Integral Equations of the Second Kind*, Cambridge University Press, 1997.
12. K. Atkinson and Weimin Han, *Theoretical Numerical Analysis: A Functional Analysis Framework*, 2nd edition, Springer-Verlag, New York, 2005.
13. K. Atkinson and L. Shampine, Algorithm 876: Solving Fredholm integral equations of the second kind in MATLAB, *ACM Trans. Math. Software*, **34** (2008), article #21.
14. C. Baker, *The Numerical Treatment of Integral Equations*, Oxford Univ. Press, 1977.

15. H. Bateman, *Report on The History and Present State of the Theory of Integral Equations*, Report of the 80th Meeting of the British Association for the Advancement of Science, Sheffield, 1910.

16. M. Bernkopf, The development of function spaces with particular reference to their origins in integral equation theory, *Archive for History of Exact Sciences* **3** (1966), 1–96.

17. H. Brakhage, Über die numerische Behandlung von Integralgleichungen nach der Quadraturformelmethode, *Numerische Math.* **2** (1960), 183–196.

18. H. Brakhage, Zur Fehlerabschätzung für die numerische Eigenwertbestimmung bei Integralgleichungen, *Numerische Math.* **3** (1961), 174–179.

19. J. Bramble and J. Osborn, Rate of convergence estimates for nonselfadjoint eigenvalue approximations, *Math. of Comp.* **27** (1973), 525–549.

20. H. Brunner, *Collocation Methods for Volterra Integral and Related Functional Equations*, Cambridge Univ.Press, 2004.

21. H. Bückner, *Die Praktische Behandlung von Integral-Gleichungen*, Springer Verlag, 1952.

22. G. Chandler, *Superconvergence of Numerical Solutions to Second Kind Integral Equations*, Ph.D. thesis, Australian National University, Canberra, 1979.

23. F. Chatelin, *Spectral Approximation of Linear Operators*, Academic Press, 1983.

24. F. de Hoog and R. Weiss, Asymptotic expansions for product integration, *Math. of Comp.* **27** (1973), 295–306.

25. L.M. Delves and J. Mohamed, *Computational Methods for Integral Equations*, Cambridge Univ. Press, 1985.

26. I. Fredholm, Sur une classe d'équations fonctionelles, *Acta Math.* **27** (1903), 365–390.

27. I. Graham, *The Numerical Solution of Fredholm Integral Equations of the Second Kind*, Ph.D. thesis, Univ. of New South Wales, Sydney, 1980.

28. L. Greengard and V. Rokhlin, A new version of the fast multipole method for the Laplace equation in three dimensions,. *Acta Numerica 1997*, Cambridge Univ. Press, Cambridge, 229–269.

29. C. Groetsch, *The Theory of Tikhonov Regularization for Fredholm Integral Equations of the First Kind*, Pitman, 1984.

30. W. Hackbusch, Die schnelle Auflösung der Fredholmschen Integralgleichungen zweiter Art, *Beiträge Numerische Math.* **9** (1981), 47–62.

31. W. Hackbusch and Z. Nowak, On the fast matrix multiplication in the boundary element method by panel clustering, *Numer. Math.* **54** (1989), 463–491.

32. W. Hackbusch, *Integral Equations: Theory and Numerical Treatment*, Birkhäuser Verlag, Basel, 1995.

33. D. Hilbert, *Grundzüge einer allgemeinen Theorie der linearen Integralgleichungen*, Teubner, Leipzig and Berlin, 1912.

34. F. Hildebrand, *Introduction to Numerical Analysis*, McGraw-Hill, 1956.

35. V. Ivanov, *The Theory of Approximate Methods and Their Application to the Numerical Solution of Singular Integral Equations*, Noordhoff, 1976.

36. M. Jaswon and G. Symm, *Integral Equation Methods in Potential Theory and Elastostatics*, Academic Press, 1977.

37. L.V. Kantorovich, Functional analysis and applied mathematics, *Uspehi Mat. Nauk* **3** (1948), 89–185. Translated from the Russian by Curtis Benster, in *NBS Report 1509* (ed. by G. Forsythe), 1952.

38. L. V. Kantorovich and V. I. Krylov, *Approximate Methods of Higher Analysis*, 3rd edition, Interscience Pub., 1964. Translated from the Russian 3rd edition of 1950 by Curtis Benster. The 2nd edition appeared in 1941.

39. L.V. Kantorovich and G. Akilov, *Functional Analysis in Normed Spaces*, 2nd edition, Pergamon Press, 1982. Translated from the Russian by Curtis Benster.

40. M. Krasnoselskii (1964), *Topological Methods in the Theory of Nonlinear Integral Equations*, Pergamon Press.

41. R. Kress, *Linear Integral Equations*, 2nd ed., Springer-Verlag, 1999.

42. P. Linz, *Analytical and Numerical Methods for Volterra Equations*, SIAM Pub., 1985.

43. A. Lonseth, Approximate solutions of Fredholm-type integral equations, *Bull. AMS* **60** (1954), 415–430.

44. A. Lonseth, Sources and applications of integral Equations, *SIAM Review* **19** (1977), 241–278.

45. I. Mysovskih, Estimation of error arising in the solution of an integral equation by the method of mechanical quadratures, *Vestnik Leningrad Univ.* **11** (1956), 66–72.

46. B. Noble, A bibliography on : "Methods for solving integral equations", Tech. Rep. **1176** (author listing), **1177** (subject listing), *U.S. Army Mathematics Research Center*, Madison, Wisconsin., 1971.

47. B. Noble, Error analysis of collocation methods for solving Fredholm integral equations, in *Topics in Numerical Analysis*, ed. by J. H. Miller, Academic Press, 1973, 211–232.

48. E. Nyström, Über die praktische Auflösung von Integralgleichungen mit Anwendungen auf Randwertaufgaben, *Acta Math.* **54** (1930), 185–204.

49. C. Pozrikidis, *Boundary Integral and Singularity Methods for Linearized Viscous Flow,* Cambridge Univ. Press, Cambridge, 1992.

50. J. Phillips, The use of collocation as a projection method for solving linear operator equations, *SIAM J. Num. Anal.* **9** (1972), 14–28.

51. P. Prenter, A method of collocation for the numerical solution of integral equations of the second kind, *SIAM J. Num. Anal.* **10** (1973), 570–581.

52. G. Richter, On weakly singular Fredholm integral equations with displacement kernels, *J. Math. Anal. & Applic.* **55** (1976), 32–42.

53. C. Schneider, Product integration for weakly singular integral equations, *Math. of Comp.* **36** (1981), 207–213.

54. I. Sloan, Improvement by iteration for compact operator equations, *Math. Comp.* **30** (1976), 758–764.

55. I. Sloan, Error analysis of boundary integral methods, *Acta Numerica* **1** (1992), 287–339.

56. A. Spence, *The Numerical Solution of the Integral Equation Eigenvalue Problem*, Ph.D. Thesis, Oxford University, 1974.

57. E. Stephan and W. Wendland, Remarks to Galerkin and least squares methods with finite elements for general elliptic problems, *Manuscripta Geodactica*, **1** (1976), 93–123.

58. F. Stummel, Diskrete Konvergenz Linearer Operatoren I, *Math. Annalen* **190** (1970), 45–92.

59. G. Vainikko, On the speed of convergence of approximate methods in the eigenvalue problem, *USSR Comp. Math. & Math. Phys.* **7** (1967), 18–32.

60. G. Vainikko, The compact approximation principle in the theory of approximation methods, *USSR Comp. Math. & Math. Phys.* **9** (1969), 1–32.

61. G. Vainikko, *Funktionalanalysis der Diskretisierungsmethoden*, Teubner, Leipzig, 1976.

62. W. Wendland, Die Behandlung von Randwertaufgaben im $\mathbf{R}^3$ mit Hilfe von Einfach- und Doppelschichtpotentialen, *Numerische Math.* **11** (1968), 380–404.

63. W. Wendland, Boundary element methods for elliptic problems, in *Mathematical Theory of Finite and Boundary Element Methods*, by A. Schatz, V. Thomée, and W. Wendland, Birkhäuser, Boston, 219–276, 1990.

64. H. Wielandt, Error bounds for eigenvalues of symmetric integral equations, *Proc. AMS Symposium Appl. Math.* **6** (1956), 261–282.

65. G. M. Wing, *A Primer on Integral Equations of the First Kind*, SIAM Pub., 1991.

66. A. Young, The application of approximate product integration to the numerical solution of integral equations, *Proc. Royal Soc. London* **A224** (1954), 561–573.

Memoires on building a general purpose numerical algorithms library

Brian Ford, OBE

Founder Director, NAG Ltd
Ramsden Farm House
High Street, Ramsden
Oxon OX7 3AU, England
brian.ford9@btinternet.com

1 Introduction

A desire to build a General Purpose Numerical Subroutine Library is not a wish most people are born with! To do it at least twice is even more unusual. Numerical Algorithm Libraries have taken much of my life – and I don't regret it! The driving force was to develop a selection of software solvers that enabled scientists and engineers to solve their problems, invariably expressed as equations, in some area of Numerical Mathematics or Statistics. In many instances this would be addressing the mathematical models underlying their research field for the first time with the realistic hope of computing predicted results.

At the outset the aim was to build a quality facility –what quality means will become clear– and the library collection would be carefully selected to meet the needs of its users over the whole field of numerical mathematics and statistics. It was not a subject library (e.g. ODEs) nor a topic library(e.g. quantum chemistry) nor a numerical analysis research group collection(e.g. the Harwell Library) but a general purpose library for University users. So in the late 1960s and early 1970s we asked our university computing users what numerical areas they needed solvers for. Then set out to discover what numerical subject areas had reasonable and useful algorithms available. From the beginning we were choosing the best – not simply taking what was available. Everyone thought we were being idealistic, but agreed to try, to help. The Library was not to be a haphazard collection, but an integrated, complete, consistently organised, well-documented and thoroughly tested selection of procedures for scientific computing.

The Result was that the NAG Library was built. Much science and engineering was completed. (We know that at least thirty Nobel Prizes were won using the Library.) There was a tremendous stimulation of Numerical Analysis and Computational Statistics, with many algorithms invented.

2 Prelude – the pre-NAG days

I solved equations using hand machines (Brunsvigas) in afternoon practicals as part of my Maths degree at Imperial College (1959-1962). I actually preferred

Quantum Mechanics but became involved with numerical computing software for the first time in 1967. I read the seminal paper [16] by Francis on the QR algorithm and started a software implementation of it. I was offered a joint post between Maths and the Computing Centre at Nottingham University. I set up a Numerical Advisory Service for the whole university in the Computing half of the post. This involved collecting and building a numerical algorithms library, preparing the documentation for user advice and creating an office where the users could come and consult.

Nottingham University had an ICL KDF-9. Its design was based on Pilot ACE, the brain child of Alan Turing, and built at NPL by Turing and then Jim Wilkinson and colleagues. The KDF-9 had a 48 bit word, hardware double-precision inner product and true rounding. The university computing resources were saturated and had to be strictly rationed. Looking to achieve improved use of the facilities, through the use of better numerical methods, I invited Jim Wilkinson and Leslie Fox to come and lecture to our users at Nottingham. To me these were the absolute Gods of numerical computing. Wilkinson had published the *Algebraic Eigenvalue Problem* in 1965 [35]. He came first and attracted 150 people to each of his two lectures. On his visit I showed Jim my QR implementation. "Do you want the good news or the bad news first" he asked. "The Good". "It's not bad". "And the bad news?" I asked. "I'm publishing my QL with Christian Reinsch in Numerische Mathematik next week!".[1] This was the beginning of my friendship with Jim, which lasted until his untimely death in 1986. Leslie Fox sent David Mayers. David also gave two excellent lectures.

These were amazing times. I built a numerical algorithms library for KDF-9. It contained over a hundred routines. It had good numerical linear algebra and random number generators, spotty coverage in non-linear optimisation, quadrature and ODEs, and one curve and surface fitting routine. Our sources of material included routines from the National Physical Laboratory and optimisation codes from Harwell. Inevitably I wrote some myself. We did some serious testing to aid reliability and prepared our first user documentation, in a distinct house style. The take-up of the material was instantaneous, by tens of people, stopping me everywhere I went. I remember going to the Playhouse and never getting back into the theatre after the interval to see the second half of the play. They were users from all over the University, delighted to be addressing their research mathematical models for the first time. Individuals were asking for specific solvers for their particular research problem. I fed queries on to NPL, Harwell and Manchester.

I also went to other universities with KDF-9s, which had active numerical analysis groups. Linda Hayes worked for Lesley Fox, and ran the user advisory desk in Oxford. Adrian Hock and Shirley Lill were in Leeds. Shirley was doing a PhD in non-linear optimisation with Roger Fletcher. None had particularly developed numerical libraries. Each had a subject library. Uniquely Manchester University had an Atlas computer. Joan Walsh had developed ODE routines

---

[1] See [2].

for local use. All of the machines ran 24/7 with demand for larger computing resources, and machine time was consistently rationed.

3   Announcement of the ICL 1906A

In November 1969 the UK Government Computer Board for Universities and Research Councils announced the provision of ICL 1906As for Oxford, Birmingham, Leeds, Manchester and Nottingham Universities and the Science Research Council Chilton Atlas Laboratory. The machine was described as so powerful that thirty per cent of its computing resources could be given over to its operating system George 3. Nottingham University was delighted to be included in such an august list. Eric Foxley, our Computing Centre Director, immediately asked me to build a numerical library for the 6A, which didn't come with one. "No" was my immediate response. I had built one, that was sufficient for any mathematician. I wanted to complete my PhD and research and teach in Quantum Mechanics. "But Brian, it's your duty. Look at all the people you've helped already. You'll be letting them all down, after creating such a community". I went away to think about it.

4   Birth of the NAG Library – built collaboratively

Lying in a bath one Saturday in February 1970, in our village of Bunny, I had a half idea. Why not build the Library collaboratively, involving the individuals I already know, or know of, in the other centres receiving 6As at the same time? I prepared for and convened a meeting, at Nottingham University, on the 13th May 1970. Linda Hayes came from Oxford, Shirley Lill and Adrian Hock came from Leeds, Joan Walsh came from Manchester and Bart Fossey came from the SRC Chilton Atlas Centre. At that inaugural meeting we agreed to build collaboratively a numerical algorithms library, in Algol 60 and Fortran, for the ICL 1906A [12]. We had resources in our computing centres which we could direct and use. Each of us was strong minded, and hard working. We led from the front and individually undertook the technical work and organisation ourselves.

The design principles for the Library were quickly agreed. The base of the Library was algorithms, individually selected and coded in both languages. The user documentation was as equally important as the source code of the routines. There would be an example program in the user documentation showing the use of each routine. Each routine would have a stringent test set to establish its correct functioning and the reasons for its selection. When an error is found in the library software or documentation, it would be corrected as soon as possible and users would be notified of its existence in the interim.

The Library was created for our University Users. We knew that the spectrum of users was wide, in terms of subject background, knowledge of computing, knowledge of numerical analysis. Formally each user had an equal claim on the computing service and its resources. Everything had to be designed to cover their range of ability, requirement and demand.

Each user had a mathematical problem to solve which fitted into one (or more) categories of numerical mathematics or statistics. The Library classification was by these areas of numerical mathematics and statistics. Most users were familiar with the vocabulary and language of their research area. Many needed support with the vocabulary and language of the mathematics of their study (e.g. Professor Sir Peter Mansfield; Nuclear Magnetic Imaging; Eigenvalue Problem). This was a major factor in preparing the user documentation.

5   Selection of library contents

The selection of the Library contents was wholly based on finding algorithms that satisfied user needs. We sought from the outset to establish an interaction between users and algorithm developers. Previously users simply accepted gratefully algorithms developed by analysts. We worked to identify user's actual needs (particularly unsolved needs) and used our contacts with numerical analysts to solve them. There were secondary factors in content selection. We recognised early that algorithm developers in one numerical area would need to utilise algorithms developed in another, and that this would affect the algorithm selected and the interface provided. For example algorithms in numerical linear algebra would be required in many parts of the Library, and there would be cross-calling to other areas too.

At our first meeting we drew up a list of preferred characteristics in algorithms, in order of importance:

1. Usefulness
2. Robustness
3. Numerical Stability
4. Accuracy
5. Speed.

(Robustness was the ability of an algorithm to fail gracefully, delivering an error message.) Usefulness was far too vague, so quickly the characteristics became:

1. Stability
2. Robustness
3. Accuracy
4. Adaptability (Trustworthiness)
5. Speed.

Speed was always last. There was too much fast rubbish about! Later Professor Christian Reinsch told Dr. Seymour Cray that his Cray 1 computer "was unsafe at any speed – because of its dirty arithmetic!" The reality was that some of our early selected algorithms had few or none of the properties. We were pressing to satisfy as best we could our user's needs so we had to include inadequate tools. We sought and seek to replace out-moded material at the earliest opportunity. The pursuit of our characteristics was a proper ideal. Library contents would be updated in an ordered, notified and consistent way.

6   Library contents

At our first meeting we also selected the first numerical and statistical areas which were to be our chapter topics, in an agreed order of priority.

1. Linear Algebra	6. Simulation
2. Statistics	7. ODEs
– Statistical Routines	8. Roots
– BMD Programs	9. Approximation Theory
3. Non-linear optimisation	10. Quadrature
4. Special Functions	11. Interpolation
5. Sorting	12. Non-linear equations

A first set of general comments on our initial, perhaps naïve decisions are appropriate. We had no long-term understanding of what we were getting into.

Clearly the numerical linear algebra came first. We stumbled over the statistics. We had a continuous argument over many years, which intensified, as to whether we should include major chapters of statistics in the Library, or direct users to employ statistical packages. We learnt that both approaches were essential. We were also slow to get the first statistical routines into use. Arguably the sorting chapter should not have been included (although it has always been heavily used). Random numbers have always been vital to the Library (described above as "simulation"). We failed to grasp the PDE nettle. It would have been the most challenging of all areas. Our failure to include PDEs set back their library availability for years. ODEs were of much greater importance than their initial placing suggests. This wrinkle in our planning was corrected by the prompt response of our ODE colleagues in Manchester and elsewhere [33]. A number of other areas were in their numerical infancy. At that time linear programming was seen as a package activity. It was good from the beginning we sought to include routines. Graphics and Visualisation deserved early attention, which they didn't get.

After some initial research we recognised four categories of algorithmic coverage for the mathematical areas addressed (26 of them!). Each chapter fell into one of

General	e.g. F02 (Eigenvalue problems)
Patchy	e.g. D02 (ODEs), E04 (non-linear optimisation)
Spotty	e.g. E02 (curve and surface fitting)
None	e.g. D05 (integral equations).

Our aim over the years was to move all chapters into general, as research and resources permitted. The Classification we chose to name the chapters was the Modified Share Classification Index. This choice was hard fought and would take of paper of itself.

Number of Routines per Chapter of the Early Libraries							
	Chapter Names	Marks of Library					
		1	2	3	4	5	6
A02	COMPLEX ARITHMETIC	—	3	—	—	—	—
C02	ZEROS OF POLYNOMIALS	—	—	—	2	—	—
C05	ROOTS OF TRANSCENDENTAL EQNS	3	2	—	—	—	—
C06	SUMMATION OF SERIES	—	3	—	1	—	1
D01	QUADRATURE	—	2	4	—	3	—
D02	ODEs	2	2	2	—	1	—
D04	NUMERICAL DIFFERENTIATION	—	—	—	—	1	—
D05	INTEGRAL EQUATIONS	—	—	—	—	1	—
E01	INTERPOLATION	2	—	2	—	—	—
E02	CURVE AND SURFACE FITTING	1	—	—	—	5	3
E04	MIN (MAX) A FTN	4	4	—	5	2	23

Marks are releases of the Library. The addition of contents is cumulative (except where stated otherwise). We had some ODE routines from the beginning. The early non-linear optimisation routines were not modularised. At Mark 5 and 6 modularly constructed routines cross-calling linear algebra modules were introduced.

Number of Routines per Chapter of the Early Libraries							
	Chapter Names	Marks of Library					
		1	2	3	4	5	6
F01	COMPLEX ARITHMETIC	2	29	11	0	2	6
F02	EIGENVALUES AND EIGENVECTORS	10	8	3	—	3	1
F03	DETERMINANTS	4	4	2	—	—	1
F04	SIMULTANEOUS LIN. EQNS	—	13	2	3	1	1
F05	ORTHOGONALISATION	—	—	—	—	1	1
G01	SIMPLE CALLS ON STATS DATA	—	—	—	3	—	—
G02	CORRELATION AND REGRESSION ANL	—	—	—	25	—	—
G04	ANL OF VARIANCE	—	—	1	—	—	—
G05	RANDOM NUMBER GNRTS	24	—	2	—	—	26

First the eigenvalue routines (F02) were introduced and then the solution of simultaneous linear equations (F04). Except for the Random Numbers the introduction of statistical routines was very slow.

Number of Routines per Chapter of the Early Libraries							
Chapter Names		**Marks of Library**					
		1	2	3	4	5	6
H	OPERATIONS RESEARCH	2	—	2	2	—	—
M01	SORTING	16	—	4	—	—	—
P01	ERROR TRAPPING	1	—	—	—	—	—
S	APPROX OF SPECIAL FTNS	8	1	3	8	7	—
X01	MATHEMATICAL CONSTANTS	—	—	—	—	2	—
X02	MACHINE CONSTANTS	—	—	—	—	9	—
X03	INNERPRODUCTS	—	—	—	—	2	—
	TOTALS	**82**	**71**	**36**	**49**	**40**	**64**

A fabulous effort by Dr Lawrie Schonfelder of Birmingham University created a methodology for the preparation and generation of portable special function routines (S chapter) [32].

Over 340 new routines were prepared and included in the first six years (and first six releases) of the Library.

7   Comments on the chapter contents developed

C05 Solution of Polynomial equations. This was a challenge area amongst numerical analysts but infrequently used.

C06 There was a significant demand for Fast Fourier Transforms. Work here led to the enlargement of our concept of algorithm.

D01 Quadrature was a specialist demand area. This led to an amazing contribution group centred on Queens University Belfast and excellent collaboration with the University of Leuven. Led by O'Hara and Patterson at different times, the group met at Queen's University, Belfast throughout the troubles in Northern Ireland and contained within it a complete political spectrum of views. Quadrature was the only topic we could all talk about. There was also an exemplary collaboration with de Doncker and Piessens centred on QUADPACK [29].

D02 After outstanding initial contributions in ODEs, specifically from Manchester, the team there led a world-wide collaborative group, with original work completed in shooting methods and the method of lines (see the case study below).

D05 Initially there was no software in the Solution of Integral Equations. Relations grew with George Symm and his colleagues at NPL [22].

E02 My own puny efforts using a mini-max polynomial to fit a set of points encouraged Geoff Hayes and Maurice Cox and their colleagues at NPL to take on leadership of the chapter (see the case study below).

E04 Early routines included the ubiquitous Simplex Method of Nelder and Mead [28], and a conjugate direction algorithm of Fletcher and Powell [11]. The contact with John Nelder led to much joint work and projects in Computational Statistics during the next thirty years. Ultimately, Walter Murray and Philip Gill of NPL co-ordinated the contribution [17].

F02 and F04 The numerical linear algebra chapters enjoyed the great input and influence of Jim Wilkinson and his colleagues. Linda Hayes and her colleagues completed an excellent translation of the Handbook Algol 60 codes [36] into Fortran, first for the eigenvalue problem and then for simultaneous linear equations. We also developed invaluable relations with the EISPACK team and other colleagues at Argonne National Laboratory (see the case study below).

G04 This was the only area where the offered software failed to meet our own agreed standards. This delayed availability of software for two years and exacerbated the arguments between subroutine and package exponents within the contribution group.

G05 The Random number package by Henry Neave was a major building block of the early libraries. It was replaced after four years by work spearheaded by Nick MacLaren [25–27].

H The Programming chapter had a slow start, initially including the LP simplex method [6] and a quadratic programming solver subject to linear constraints.

P01 The systematic error reporting from the Library routines, fundamentally dependent on the Library Documentation, was one of the immediate successes of the NAG Library, creating great user confidence.

S Lawrie Schonfelder (and colleagues) had the awesome responsibility of generating full machine precision approximations for the commonly used special functions (defined by the "Handbook of Special Functions" by Abramowitz and Stegun [1]). This he achieved progressively over the early marks of the Library [31].

X02 Once the issue of the portability of numerical software was grasped, the study of machine characteristics and their parameterisation in numerical software became a research study. Finally a set of parameters was chosen for the Library [13].

8   Chapter subdivisions

Within each overall chapter we were looking for subdivisions into different problem types. This was important when selecting algorithms for inclusion in the Library. In linear algebra we had eigensolution and simultaneous equations, and then in each section real symmetric, real, hermitian, and complex matrices so that linear algebra problems truly subdivided into individual problem types requiring an identifiable individual algorithm. Problem types in non-linear optimisation could only be partially identified.

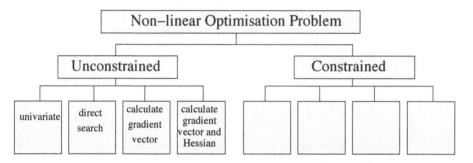

Fig. 1 Classification of the chapters, optimisation example.

Beginning with the separation into unconstrained and constrained, we could then classify the problem type by the information available to compute the objective function. But amongst all problems solved by direct search techniques, could not subtypes be found, defined by the mathematical characteristics of the objective function? And then select the best algorithm for each subtype? That was the aim.

9   Library contribution

In building the Library we were and are fundamentally dependent on our individual Chapter Contributors. The selection of algorithms for inclusion in the Library is a fundamental task with far-reaching consequences and implications. Initially we divided the responsibilities between the co-operating centres as shown in the table below, on the basis of local expertise and the desire to achieve complete initial coverage of the chosen material. In some areas chapter leadership remained with the original contributing centre whilst for others it soon passed to more expert hands.

	Chapter Contents	Collating Centre	Test Centre
F01-F05	Linear Algebra	Oxford	Nottingham
G01-G04	Statistics		
	–Statistical routines	Manchester	Birmingham
	–BMD programs	Nottingham	Birmingham
E04	Non-linear Optimisation	Leeds	Manchester
S	Special Functions	Birmingham	Manchester
M	Sorting	Birmingham	Manchester
G05	Simulation	Nottingham	Manchester
D02	ODEs	Manchester	Oxford
C05	Roots	Atlas	Leeds
E02	Approximation Theory	Nottingham	Leeds
D01	Quadrature	Atlas	Oxford
H	Programming	Manchester	Birmingham
E01	Interpolation	Nottingham	Leeds
C05	Non-linear Equations	Leeds	Nottingham

The activity in linear algebra worked like a dream, with close collaboration between Linda Hayes and colleagues in Oxford and Jim Wilkinson and his group at NPL. In Statistics we soon dropped the BMD programs. We were so grateful for so many wonderful colleagues (particularly Linda Hayes, Shirley Lill and Joan Walsh) who simply put their heads down, and worked!

As the success of the activity became clear a series of workshops were organised, following which individual chapter groups were formed and group leaders recognised (or particular research groups). Each chapter contributor defined the sub-structuring of their chapter and selected the algorithmic contents accordingly. As each was an expert numerical analyst (or statistician) in the particular area, they knew the currently available algorithms, what was inadequate, who was working in the given sub-fields and what was involved in developing new algorithms to fill holes. These contributors were and are voluntary and unpaid. Their interest in involvement was to drive their area forward, broaden their own contacts and expertise, see their work used and the increased reference and published use of their algorithms.

10   Three chapter case studies

10.1   Numerical linear algebra

The numerical linear algebra was crucial to the Library activity. Oxford completed the translation of the Handbook for Linear Algebra Algol 60 codes [36] into Fortran. The NATS EISPACK project at Argonne National Laboratory did the same [32]. Jim Wilkinson gave his blessing to the NAG Library project and committed to work with us. This he did for the rest of his life.

Jim appreciated the difference between Numerical Analysis, Algorithm Design and Software Implementation, and valued all three. He believed in cooperation and collaboration, and valued critical analysis of his own work. We shared with him a philosophy of excellence, in every phase of numerical software creation. Jim introduced me, and other members of NAG to the group at Argonne, to EISPACK [32] and LINPACK [8], and to other numerical linear algebraists in North America and Europe. The reaction everywhere was "if it's good enough for Jim, it's good enough for me!"

The Linear Algebra Working Party, chaired by Jim, met regularly from 1972 at NPL. It worked filling in the holes in the numerical linear algebra offering, planned the inclusion and development of the BLAS [9, 10, 19] in the Library, prepared input from and for the LINPACK project, collaborated with John Reid (a member of the working party) in the inclusion of sparse matrix software and regularly updated the existing Library contents.

Jim gave us his backward error analysis [34], his understanding of algorithm stability, the Handbook in Linear Algebra with Reinsch [36] and input designing basic linear algebra modules for use throughout the Library. Wilkinson was a welcoming and friendly man with everyone, a natural collaborator.

10.2   Curve and surface fitting, and interpolation

Geoff Hayes took charge of the E02 contribution. First the polynomial and spline routines already in Algol 60 were translated into Fortran. Their software was built on Clenshaw's work in Chebyshev polynomials [3] and Wilkinson's linear algebra [36]. All contributions were supported by floating-point error analyses (except for least squares polynomials!). Later there were new algorithms for B-splines [5] mirroring the work of de Boor [7], and basic algorithms for differentiation and integrating polynomials and splines. The contributions were validated by Ian Gladwell whom Maurice Cox described as "a hard task Master". They had a number of fiery meetings and correspondence. His mature view (twenty years later!) was that "I'm glad we took the criticism. The resulting code and documentation was much improved".

10.3   Ordinary differential equations

Joan Walsh was the only experienced numerical analyst in the NAG founding Group. She took responsibility for the ODEs and gave us much excellent software from the inception of the Library. Exhaustive testing was a great feature of the Manchester Group (George Hall, Ian Gladwell and David Sayers with Joan). At the first meeting on ODE contents at Manchester, during a discussion of Shooting Methods for Boundary Value problems, Fox made clear his great distrust of computers, and why fundamentally he would not use them!

An ODE meeting was arranged at Argonne National Laboratory hosted by George Byrne and including Bill Gear, Alan Hindemarsh, Tom Hull, Fred Krogh, Larry Shampine and Marilyn Gordan. It discussed and largely agreed issues of shared technical vocabulary, testing regimes and identification of problem types. The ODE groups at Manchester and Toronto had close technical relations, particularly on testing.

11   Types of Library Software

The majority of user requirements were met with three types of software.

Type	Function	Example
Problem solver	one routine call to solve a problem	Solution of set of simultaneous, real linear equations
Primary routine	each routine contains one major algorithm	LU factorization
Basic module	basis numerical utility designed for contributors use	extended precision inner-product

12   Library construction and operation — the NAG Library Machine

For the outline of the NAG Library Machine see Figure 2.

From the outset preparation of the Library, taking it through all its stages to its ultimate use, by users, was a substantial collaboration by tens, then quite quickly hundreds of people, the vast majority of whom worked together completely voluntarily, without payment. I called this team and process "The NAG Library Machine" [14].

13   Issues of numerical software portability

Within weeks of distributing our first NAG Library, the NAG ICL 1906A Mark 1 Library (on 30th September 1971) Manchester Regional Computing Centre (which had both an ICL 1906A and CDC 7600 computers) was asking for a "7600 copy". Cambridge University soon joined asking for an IBM version. These machines had different bases of floating-point representation, different word lengths and different overflow/underflow thresholds. We were immediately immersed in the twin problems of algorithmic and software portability.

Recognising and ultimately solving these issues of portability was one of the greatest achievements of the NAG Library project. It was a real contribution to scientific and technical computing [18]. We also recognised the problems of

The "Machine"

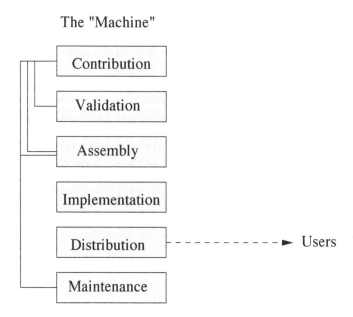

Fig. 2 The NAG Library Machine

computing language dialects and overcame these by using strict language subsets [30]. Our aim was routines, running to prescribed efficiency and accuracy on all the machines to which we took the Library, with the routines in the many versions virtually unchanged [15]. We needed and developed adaptable algorithms realised as transportable (later called portable) subroutines.

14   NAG Library Conceptual Machine

Our Contributors wished to write a single routine, which would be tailored automatically to perform to required accuracy and efficiency on every computing configuration to which the routine was carried. As a basis for the development of all NAG software we defined a simple model of the numerical and related features of any computer, relevant to the execution and performance of the software. We parameterised this model, and developed all numerical software based on this model [13].

So the conceptual machine was described in terms of these parameters, for example overflow threshold, base of floating-point number representation – sradix is 16 for the IBM 360, 2 for the CDC 7600. Hence the Contributor designed their algorithm and wrote the routine using the parameters of the conceptual machine. The algorithm and routine were then tailored for a particular machine by the choice of values for the parameters. The Contributor provided an implementation test for each routine to demonstrate its correct working in each separate environment.

15   Models of portability for numerical software

There were Numerical Software Portability Wars in the early and middle 1970s. One group sought portability through use of a single commonly used high quality computer arithmetic. Another group wanted software portability so that user programs could move easily between different computing configurations. Both groups were correct in their desires and objectives and were ultimately successful. Cody, Kahan and colleagues prepared the IEEE Arithmetic Standard [23, 21] and saw it adopted by the majority of world computing systems (particularly those for scientific computing). We, as part of the software group, knew we had succeeded when Kahan said to me in 1976 "Brian, we're all writing portable numerical software now!"

16   NAG Library Manual

For NAG the user documentation was and is always as important as the software it supports. It needs to cater for different levels of mathematical and computing knowledge and experience in its users. Further the documentation mirrors the issues addressed by the software, handling the evolution of Library contents (updating), supporting the software on many different computer ranges and has the same chapter structure in the manual as in the Library. The challenge of "portable documentation/information for portable software" was only finally met by the NAG Library Engine in the late 1990s.

To meet these requirements the manual is document-based. Each chapter starts with a substantial introduction with four distinct elements:

1. Background information on subject area
2. Recognition of problem type
3. Recommendation on choice and use of routine
4. Flow chart guide to selection of routine.

There is then a separate routine document for each routine in the chapter, which describes all aspects of the routine and its use under 13 headings, including an example program with input data and results. Each routine and its routine document have the same name.

In its day the NAG Manual was an international success. Later it was felt to be too big, even cumbersome.

17   Software testing

There were and are at least six different types of testing for Library Development.

1. Algorithm selection (Contributor and Validator)
2. Stringent testing (assembly)
3. Language standards compliance (assembly)
4. Implementation testing (Implementation)

5. Example programs

6. Automatic results comparison.

All are important in different phases of Library preparation. Along with the standards, the testing is what brought the Library its high quality (and high quality name).

Each contributor has their own independent test suite for algorithm testing and selection. Each validator has their own set too! The codes for the stringent testing of the software are selected by the contributors, and as a collection of software is bigger than the Library code itself! Automatic results comparison became an art and a science in its own right.

18   Algorithm testing

The algorithm testing by each contributor was paramount.

Linear algebra. Wilkinson set the standard for the activity with an error analysis for each algorithm including detailed stability analysis, and studying the effects of overflow and underflow. He used these to choose Householder transformations for the QL algorithm. Brian Smith in EISPACK used backward error analysis tests to check for algorithm portability and software consistency.

Special functions. Kuki and Cody did comparable analysis preparing approximations to elementary and special function. There were no backward errors and Cody could not compute forward errors (without a massively expensive multiprecision facility-which would have blown models of portability.) Cody "tested the corners of his special functions using his Elefunt system". He was the first to use mathematical identities to check the correctness of computed results [4].

Quadrature. Valerie Dixon noted at the Software for Numerical Mathematics Conference at Loughborough in 1972 "Now that many good automatic routines are available the problem of comparing them all will become extremely difficult unless the procedure for testing each new routine is standardised". As the field developed, Lyness showed the ultimate ineffectiveness of this approach. He did similar work to Cody, which resulted in showing the gaps and limitations in evaluating algorithms in quadrature. Lyness commented "it is easier to develop 3 new quadrature algorithms than establish that your existing algorithm is giving "correct" results". "You could always construct a function that would defeat a given quadrature rule". These experiences led Lyness to his seminal idea of creating performance profiles for quadrature algorithms [24], later applied in other fields of numerical analysis. This underlines the need for care and artful judgement in specifying for which integrand type a particular algorithm is used.

ODEs. The careful testing by Walsh and Gladwell in selection of their first general-purpose initial value solver was mentioned above. Their reference was to "Comparing numerical methods for ODEs", Hull, Enright et al [1972] [20]. The work continues today. The ODE community has maintained a tradition of careful public analysis of competing algorithms.

19   Validation and library assembly

Validation involves the certification of algorithm selection and user documentation of a particular chapter contribution. The activity is invariably based on an additional test suite. Validation of another's software and documentation is still a great way to make friends!

Library assembly is the work of many individuals. It is the point at which the software instantiation of a particular algorithm is thoroughly tested. It is where inconsistency and confusion of agreed standards is removed, and common standards in software and documentation enforced. Hence any diversity in algorithm design and software development is "ironed out" and a unified style created. Where-ever possible, standards are machine proven, using specific software tools created for these purposes. Inevitably some checking is by hand for example checking that the relevant chapter design is being followed and that subroutine interfaces follow general Library structure.

20   Software certification

A sophisticated set of software tools, to insure consistency and reliability of the contributed software was prepared, or "borrowed" to complete the certification process. These tools, together with the language compilers were and are:

1. Diagnosing a coding error (algorithmic or linguistic)
2. Altering a structural property of the text (e.g. ordering of Fortran statements)
3. Standardising appearance
4. Standardising nomenclature
5. Conducting dynamic analysis (e.g. branch analysis)
6. Ensuring adherence to language standards
7. Changing operational properties (e.g. generate double-precision version)
8. Coping with arithmetic, dialect etc. between computing configurations.

Then there was an initial implementation check on ICL 1900(48 bit), IBM (4/8 byte) and CDC 7600(60 bit) with stringent, implementation and example programs to certify accuracy, efficiency and effectiveness.

The Mark 6 Library = Mark 5 Library + 64 new routines!

21   Implementation and distribution to sites

At Mark 5 there were 44 different implementations of the Fortran Library and 16 in Algol 60. Currently at Mark 21 in 2008 there are 78 different implementations of the Fortran Library and 39 of the equivalent C Library.

The Library in its many implementations has been distributed to thousands of different using sites during the last 37 years. A modified programme of implementation, distribution and support continues today. The Numerical Algorithms Group (NAG) is very much alive, building Libraries and serving the Scientific Computing Community.

22   Operational principles

▷	Consultation:	Library for users
		Library Machine helps all its parts
▷	Collaboration:	gives access to required resources
▷	Co-ordination:	parallel, independent development in unified structure
▷	Planning:	overall design of individual chapter contents
▷	Mechanisation:	minimises costs, whilst giving most reliable software processing

23   Conclusions

The building of the NAG Library, and similar library ventures, led to the designing of scalable numerical algorithms and portable numerical software.

With related technical developments (IEEE arithmetic, visualisation and symbolic computing) they resulted in a great flowering of Computational Science, Engineering, Economics, Finance and Medicine.

Fundamental to all of this was the underpinning Numerical Mathematics, Numerical Analysis and Numerical Software.

A similar spirit of collaboration continues in NAG today.

Acknowledgments to the hundreds of collaborators in the NAG Library project over the years, who made it all possible, and to Drs Adhemar Bultheel and Ronald Cools for such a delightful meeting in Leuven, Belgium.

References

1. M. Abramowitz and A. Stegun. *Handbook of mathematical functions with formulas, graphs and mathematical tables.* U.S. Department of Commerce, Washington, 1964.
2. H. Bowdler, R.S. Martin, C. Reinsch, and J.H. Wilkinson. The QR and QL algorithms for symmetric matrices. *Numerische Mathematik,* **11** (1968) 293–306.

3. C.W. Clenshaw and J.G. Hayes. Curve and surface fitting. *J. Inst. Math. Appl.*, **1** (1965) 164–183.

4. W.J. Cody. The construction of numerical subroutine libraries. *SIAM Rev.*, **16** (1974) 36–45.

5. M.G. Cox. The numerical evaluation of a spline from its B-spline representation. *J. Inst. Math. Appl.*, **21** (1978) 135–143.

6. G.B. Dantzig. *Linear programming and extensions.* Princeton University Press, 1963.

7. C. de Boor. Bounding the error in spline interpolation. *SIAM Rev.*, **16** (1974) 531–544.

8. J.J. Dongarra, J.R. Bunch, C.B. Moler, and G.W. Stewart. *LINPACK, users' guide.* SIAM, 1979.

9. J.J. Dongarra, J. Du Croz, S. Hammarling, and R.J. Hanson. An extended set of basic linear algebra subprograms. *ACM Trans. Math. Software*, **14** (1988) 18–32.

10. J.J. Dongarra, J. Du Croz, S. Hammarling, and I.S. Duff. A set of level 3 basic linear algebra subprograms. *ACM Trans. Math. Software*, **16** (1990) 1–17.

11. R. Fletcher and M.J.D. Powell. A rapidly convergent descent method for minimization. *Comput. J.*, **6** (1963) 163–168.

12. B. Ford. Developing a numerical algorithms library. *IMA Bulletin*, **8** (1972) 332–336.

13. B. Ford. Parameterisation for the environment for transportable numerical software. *ACM Trans. Math. Software*, **4** (1978) 100–103.

14. B. Ford, J. Bentley, J.J. Du Croz, and S.J. Hague. The NAG Library "Machine". *Software Practice and Experience*, **9** (1979) 65–62.

15. B. Ford and D.K. Sayers. Developing a single numerical algorithms library for different machine ranges. *ACM Trans. Math. Software*, **2** (1976) 115–131.

16. J. Francis. The QR transformation, Part I and Part II. *Comput. J.*, **4** (1961) 265–271 and 332–345.

17. P.E. Gill, W. Murray, S.M. Picken, and M.H. Wright. The design and structure of a Fortran program library for optimization. *ACM Trans. Math. Software*, **5** (1979) 259–283.

18. S.J. Hague, and B. Ford. Portability Prediction and Correction. *Software Practice and Experience*, **6** (1976) 61–69.

19. R.J. Hanson, D. Kincaid, F.T. Krogh, and C.L. Lawson. Basic linear algebra subprograms for Fortran usage. *ACM Trans. Math. Software*, **5** (1979) 153–165.

20. T.E. Hull, W.H. Enright, B.M. Fellen, and A.E. Sedgwick. Comparing numerical methods for ordinary differential equations. *SIAM J. Numer. Anal.*, **9** (1972) 603–637.

21. IEEE Standard for binary floating-point arithmetic. *ANSI/IEEE Std 754-1985*, 1985. http://ieeexplore.ieee.org/stamp/stamp.jsp?arnumber=30711

22. M. Jaswon and G. Symm, *Integral equation methods in potential theory and elastostatics*, Academic Press, 1977.

23. W.M. Kahan and J. Palmer. On a proposed floating-point standard. *ACM SIGNUM Newsletter*, **14** (1979) 13–21.

24. J.N. Lyness and J.J. Kaganove. A technique for comparing automatic quadrature routines. *Computer J.*, **20** (1977) 170–177.

25. M.D. MacLaren, G. Marsaglia, and T.A. Bray. A fast procedure for generating exponential random variables. *Commun. ACM*, **7**, (1964) 298–300.

26. M.D. MacLaren and G. Marsaglia. Uniform random number generators. *J. Assoc. Comput. Mach.*, **12**, (1965) 83–89.

27. G. Marsaglia, M.D. MacLaren and T.A. Bray. A fast procedure for generating normal random variables. *Commun. ACM*, **7**, (1964) 4–10.

28. J.A. Nelder and R. Mead. A simplex method for function minimization. *Computer J.*, **7** (1965) 308–313.

29. R. Piessens, E. de Doncker-Kapenga, C.W. Überhuber, and D.K. Kahaner. *Quadpack. A subroutine package for automatic integration*, volume 1 of *Springer Series in Computational Mathematics*. Springer Verlag, 1983.

30. B. Ryder. The PFORT Verifier. *Software Practice and Experience*, **4** (1974) 359–377.

31. J.L. Schonfelder. The production of special function routines for a multi-machine library. *Softw., Pract. Exper.*, **6** (1976) 71–82.

32. B.T. Smith, J.M. Boyle, B.S. Garbow, Y. Ikebe, V.C. Klema, and C.B. Moler. *Matrix eigensystem routines - EISPACK guide*, volume 6 of *Notes in Computer Science*. Springer Verlag, 1974.

33. J. Walsh. *Initial and boundary value routines for ordinary differential equations.* in Software for Numerical Methods, D.J. Evans (ed.) Academic Press pp. 177–189 (1974).

34. J.H. Wilkinson. *Rounding errors in algebraic processes*. Prentice Hall, 1963.

35. J.H. Wilkinson. *The algebraic eigenvalue problem*. Claredon Press, 1965.

36. J.H. Wilkinson and C. Reinsch. *Handbook for automatic computation. Volume II, Linear algebra*, volume 186 of *Die Grundlehren der mathematischen Wissenschaften*. Springer Verlag, 1971.

Recent trends in high performance computing

Jack J. Dongarra[1], Hans W. Meuer[2], Horst D. Simon[3], and Erich Strohmaier[4]

[1] Computer Science Department, University of Tennessee, Knoxville, TN 37996
Mathematical Science Section, Oak Ridge National Lab., Oak Ridge, TN 37831; The
University of Manchester, Manchester UK;
dongarra@cs.utk.edu
[2] Computing Center, University of Mannheim, D-68131 Mannheim, Germany ;
meuer@rz.uni-mannheim.de
[3] Lawrence Berkeley Laboratory, 50A, Berkeley, CA 94720;
simon@nersc.gov
[4] CRD, Lawrence Berkeley Laboratory, 50F1603, Berkeley, CA 94720;
estrohmaeir@lbl.gov

Abstract. In this paper we analyze major recent trends and changes in
High Performance Computing (HPC). The introduction of vector computers started the area of "Supercomputing". The initial success of vector
computers in the seventies was driven by raw performance. Massive Parallel Processors (MPPs) became successful in the early nineties due to
their better price/performance ratios, which was enabled by the attack
of the "killer-micros". The success of microprocessor based Symmetric
MultiProcessor (SMP) concepts even for the very high-end systems, was
the basis for the emerging cluster concepts in the early 2000's. Within the
first half of this decade clusters of PC's and workstations have become
the prevalent architecture for many HPC application areas on all ranges
of performance. However, the Earth Simulator vector system demonstrated that many scientific applications can benefit greatly from other
computer architectures. At the same time there is renewed broad interest in the scientific HPC community for new hardware architectures and
new programming paradigms. The IBM BlueGene/L system is one early
example of a shifting design focus for large-scale system.

Key Words: High Performance Computing, HPC, Supercomputer Market, HPC
technology, Supercomputer market, Supercomputer technology

1 Introduction

"The Only Thing Constant Is Change" – Looking back on the last four decades
this seems certainly to be true for the market of High Performance Computing
systems (HPC). This market was always characterized by a rapid change of
vendors, architectures, technologies and the usage of systems. Despite all these
changes the evolution of performance on a large scale however seems to be a
very steady and continuous process. Moore's Law is often cited in this context.
If we plot the peak performance of various computers of the last six decades in

Fig. 1 which could have been called the "supercomputers" of their time [2, 4] we indeed see how well this law holds for almost the complete lifespan of modern computing. On average we see an increase in performance of two magnitudes of order every decade.

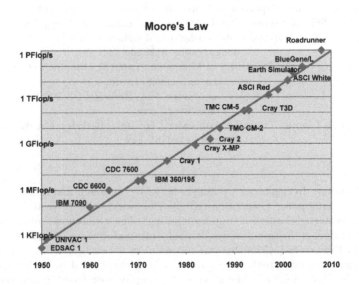

Fig. 1 Performance of the fastest computer systems for the last six decades compared to Moore's Law.

In this paper we analyze recent major trends and changes in the HPC market. For this we focus on systems, which had at least some commercial relevance. This paper extends a previous analysis of HPC market in [6]. Historical overviews with different focus can be found in [8, 9]. Section 2 summarizes our earlier finding [6]. Section 3 analyzes the trend in the first half of this decade and section 4 projects our finding into the future.

The initial success of vector computers in the seventies was driven by raw performance. The introduction of this type of computer systems started the area of "Supercomputing". In the eighties the availability of standard development environments and of application software packages became more important. Next to performance these criteria determined the success of MP vector systems especially at industrial customers. MPPs became successful in the early nineties due to their better price/performance ratios, which was enabled by the attack of the "killer-micros". In the lower and medium market segments the MPPs were replaced by microprocessor based SMP systems in the middle of the nineties. Towards the end of the nineties only the companies which had entered the emerging markets for massive parallel database servers and financial applications attracted enough business volume to be able to support the hardware development for the numerical high end computing market as well. Success in the traditional floating point intensive engineering applications was no longer sufficient for survival in the market. The success of microprocessor based SMP concepts even for the very high-end systems was the basis for the emerging cluster concepts in the early 2000s. Within the first half of this decade clusters of PC's and workstations have become the prevalent architecture for many application areas in the TOP500 on all ranges of performance. However, the Earth Simulator vector system demonstrated that many scientific applications can benefit greatly from other computer architectures. At the same time there is renewed broad interest in the scientific HPC community for new hardware architectures and new programming paradigms. The IBM BlueGene/L system is one early example of a shifting design focus for large-scale system. The IBM Roadrunner system at Los Alamos National Laboratory reached the Petaflops threshold in June 2008.

2 A short history of supercomputers

In the second half of the seventies the introduction of vector computer systems marked the beginning of modern supercomputing. These systems offered a performance advantage of at least one order of magnitude over conventional systems of that time. Raw performance was the main if not the only selling argument. In the first half of the eighties the integration of vector systems in conventional computing environments became more important. Only the manufacturers which provided standard programming environments, operating systems and key applications were successful in getting industrial customers and survived. Performance was mainly increased by improved chip technologies and by producing shared memory multiprocessor systems.

Fostered by several Government programs massive parallel computing with scalable systems using distributed memory got into the center of interest at the end of the eighties. Overcoming the hardware scalability limitations of shared memory systems was the main goal for their development. The increase of performance of standard micro processors after the RISC revolution together with the cost advantage of large scale productions formed the basis for the "Attack of the Killer Micro". The transition from ECL to CMOS chip technology and the usage of "off the shelf" micro processor instead of custom designed processors for MPPs was the consequence. Traditional design focus for MPP system was the very high end of performance. In the early nineties the SMP systems of various workstation manufacturers as well as the IBM SP series, which targeted the lower and medium market segments, gained great popularity. Their price/performance ratios were better due to the missing overhead in the design for support of the very large configurations and due to cost advantages of the larger production numbers. Due to the vertical integration of performance it was no longer economically feasible to produce and focus on the highest end of computing power alone. The design focus for new systems shifted to the market of medium performance systems.

The acceptance of MPP systems not only for engineering applications but also for new commercial applications especially for database applications emphasized different criteria for market success such as stability of system, continuity of the manufacturer and price/performance. Success in commercial environments became a new important requirement for a successful supercomputer business towards the end of the nineties. Due to these factors and the consolidation in the number of vendors in the market hierarchical systems built with components designed for the broader commercial market replaced homogeneous systems at the very high end of performance. The marketplace adopted clusters of SMPs readily, while academic research focused on cluster of workstations and PCs.

3   2000-2005: Cluster, Intel processors, and the Earth Simulator

In the early 2000's clusters built with of the shelf components gained more and more attention not only as academic research objects but also computing platforms with end-users of HPC computing systems. By 2004 these group of clusters represent the majority of new systems on the TOP500 in a broad range of application areas. One major consequence of this trend was the rapid rise in the utilization of Intel processors in HPC systems. While virtually absent in the high end at the beginning of the decade, Intel processors are now used in the majority of HPC systems. Clusters in the nineties were mostly self-made system designed and built by small groups of dedicated scientist or application experts. This changed rapidly as soon as the market for clusters based on PC technology matured. Nowadays the large majority of TOP500-class clusters are manufactured and integrated by either a few traditional large HPC manufacturers such as IBM or HP or numerous small, specialized integrators of such systems.

In 2002 a system called "Computnik" with a quite different architecture, the Earth Simulator, entered the spotlight as new #1 system on the TOP500 and it managed to take the U.S. HPC community by surprise even though it had been announced 4 years earlier. The Earth Simulator built by NEC is based on the NEC vector technology and showed unusual high efficiency on many applications. This fact invigorated discussions about future architectures for high-end scientific computing systems. A first system built with a different design focus bust still with mostly conventional off the shelf components is the BlueGene/L system. Its design focuses on a system with an unprecedented number of processors using a power efficient design while sacrificing main memory size

3.1   Explosion of cluster based system

By the end of the nineties clusters were common in academia but mostly as research objects and not so much as computing platforms for applications. Most of these clusters were of comparable small scale and as a result the November 1999 edition of the TOP500 listed only 7 cluster systems. This changed dramatically as industrial and commercial customer started deploying clusters as soon as their applications permitted to take advantage of the better price/performance ratio of commodity based clusters. At the same time all major vendors in the HPC market started selling this type of clusters fully integrated to their customer base. In November 2004 clusters became the dominant architecture in the TOP500 with 294 systems at all levels of performance (see Fig. 2). Companies such as IBM and Hewlett-Packard sell the majority of these clusters and a large number of them are installed at commercial and industrial sites. To some

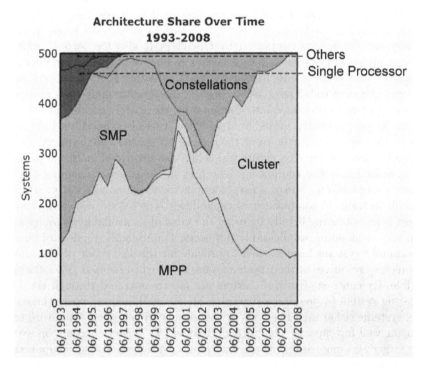

Fig. 2 Main architectural categories seen in the TOP500 (the term Constellations refers to clusters of SMPs).

extent, the reasons for the dominance of commodity-processor systems are economic. Contemporary distributed —memory supercomputer systems based on commodity processors (like Linux clusters) appear to be substantially more cost effective— roughly an order of magnitude-in delivering computing power to applications that do not have stringent communication requirements. On the other hand, there has been little progress, and perhaps regress, in making scalable systems easy to program. Software directions that were started in the early 80's (such as CMFortran and High Performance Fortran) were largely abandoned. The payoff to finding better ways to program such systems and thus expand the domains in which these systems can be applied would appear to be large. The move to distributed memory has forced changes in the programming paradigm of supercomputing. The high cost of processor-to-processor synchronization and communication requires new algorithms that minimize those operations. The structuring of an application for vectorization is seldom the best structure for parallelization on these systems. Moreover, despite some research successes in this area, without some guidance from the programmer, compilers are generally able neither to detect enough of the necessary parallelism, nor to reduce sufficiently the inter-processor overheads. The use of distributed memory systems has led to the introduction of new programming models, particularly the message passing paradigm, as realized in MPI, and the use of parallel loops in shared memory subsystems, as supported by OpenMP. It also has forced significant reprogramming of libraries and applications to port onto the new architectures. Debuggers and performance tools for scalable systems have developed slowly, however, and even today most users consider the programming tools on parallel supercomputers to be inadequate. Fortunately, there are a number of choices of communication networks available in addition; there is generally a large difference in the usage of clusters and their more integrated counterparts: clusters are mostly used for capacity computing while the integrated machines primarily are used for capability computing. The first mode of usage meaning that the system is employed for one or a few programs for which no alternative is readily available in terms of computational capabilities. The second way of operating a system is in employing it fully by using the most of its available cycles by many, often very demanding, applications and users. Traditionally, vendors of large supercomputer systems have learned to provide for this last mode of operation as the precious resources of their systems were required to be used as effectively as possible. By contrast, Beowulf clusters are mostly operated through the Linux operating system (a small minority using Microsoft Windows) where these operating systems either miss the tools or these tools are relatively immature to use a cluster well for capacity computing. However, as clusters become on average both larger and more stable, there is a trend to use them also as computational capacity servers.

3.2   Intel-ization of the processor landscape

The HPC community had started to use commodity parts in large numbers in the nineties already. MPPs and Constellations (the term Constellations refers to cluster of SMPs) are typically using standard workstation microprocessors, even so they still might use custom interconnect systems. There was however one big exception, virtually nobody used Intel microprocessors. Lack of performance and the limitations of a 32 bit processor design were the main reasons for this. This changed with the introduction of the Pentium 3 and especially in 2001 with the Pentium 4, which featured greatly improved memory performance due to its front-side bus and full 64bit floating point support. The number of systems in the TOP500 with Intel processors exploded from only 6 in November 2000 to 375 in June 2008 (Fig. 3).

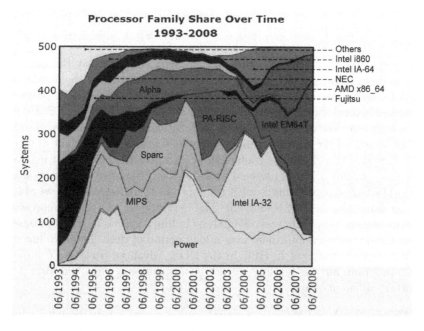

Fig. 3 Main processor families seen in the TOP500.

3.3   The Earth Simulator shock

The Earth Simulator (ES) was conceived, developed, and implemented by Hajime Miyoshi. Dr. Hajime Miyoshi is regarded as the Seymour Cray of Japan. Unlike his peers, he seldom attended conferences or gave public speeches. However, he was well known within the HPC community in Japan for his involvement in the development of the first Fujitsu supercomputer in Japan, and later on of the Numerical Wind Tunnel (NWT) at NAL. In 1997 he took up his post as the director of the Earth Simulator Research & Development Center (ESRDC) and led the development of the 40 Tflop/s Earth Simulator, which would serve as a powerful computational engine for global environmental simulation.

Prior to the ES, global circulation simulations were made using a 100km grid width although ocean-atmospheric interactive analyses were not performed. To get quantitatively good predictions for the evaluation of environmental effects may require grid width of at most 10 km or 10 times finer meshes in x, y and z directions and interactive simulation. Thus a supercomputer 1000 times faster and larger than a 1995 conventional supercomputer might be required. Miyoshi investigated whether such a machine could be built in the early 2000s. His conclusion was that it could be realized if several thousand of the most advanced vector supercomputers of approximately 10 Gflop/s speed were clustered using a very high-speed network. He forecasted that extremely high-density LSI integration technology, high-speed network (crossbar) technology, as well as an efficient operating system and Fortran compiler all could be developed within the next several years. He thought only a strong initiative project with government financial support could realize this kind of machine.

The machine was completed in February, 2002 and presently the entire system continues to be used as an end user service. He supervised the development of NWT Fortran as the leader of NWT project and organized HPF (High Performance Fortran) Japan Extension Forum, which is used on the ES. He knew that a high-level vector/parallel language is critical for such a supercomputer.

The launch of the Earth Simulator created a substantial amount of concern in the U.S. that it had lost the leadership in high performance computing. While there was certainly a loss of national pride for the U.S. not to be first on a list of the world's fastest supercomputers, it is important to understand the set of issues that surround that loss of leadership. The development of the ES represents a large investment (approximately $500M, including a special facility to house the system) and a large commitment over a long period of time. The U.S. has made an even larger investment in HPC in the DOE Advanced Strategic Computing (ASC) program, but the funding has not been spent on a single platform. Other important differences are:

- ES was developed for basic research and is shared internationally, whereas the ASC program is driven by national defense and the systems have restricted domestic use.
- A large part of the ES investment supported NEC's development of their SX-6 technology. The ASC program has made only modest investments in industrial R&D.
- ES uses custom vector processors; the ASC systems use commodity processors.
- The ES software technology largely originates from abroad – although it is often modified and enhanced in Japan. For example, significant ES codes were developed using a Japanese enhanced version of HPF. Virtually all software used in the ASC program has been developed by the U.S.

Surprisingly, the Earth Simulator's number one ranking on the TOP500 list was not a matter of national pride in Japan. In fact, there is considerable resentment of the Earth Simulator in some sectors of research communities in Japan. Some Japanese researchers feel that the ES is too expensive and drains critical resources from other science and technology projects. Due to the continued economic crisis in Japan and the large budget deficits, it is getting more difficult to justify government projects of this kind.

3.4   New architectures on the horizon

Interest in novel computer architectures has always been large in the HPC community, which comes at little surprise, as this field was borne and continues to thrive on technological innovations. Some of the concerns of recent years were the ever-increasing space and power requirements of modern commodity based supercomputers. In the BlueGene/L development, IBM addressed these issues by designing a very power and space efficient system. BlueGene/L does not use the latest commodity processors available but computationally less powerful and much more power efficient processor versions developed mainly not for the PC and workstation market but for embedded applications. Together with a drastic reduction of the available main memory, this leads to a very dense system. To achieve the targeted extreme performance level and unprecedented number of these processors (up to 212,992) are combined using several specialized interconnects.

There was and is considerable doubt whether such a system would be able to deliver the promised performance and would be usable as a general-purpose system. First results of the current beta-System are very encouraging and the one-quarter size beta- System of the future LLNL system was able to claim the number one spot on the November 2004 TOP500 list.

4 2005 and beyond

Three decades after the introduction of the Cray 1 the HPC market had changed its face quite a bit. It used to be a market for systems clearly different from any other computer systems. Today the HPC market is no longer an isolated niche market for specialized systems. Vertically integrated companies produced systems of any size. Components used for these systems are the same from an individual desktop PC up to the most powerful supercomputers. Similar software environments are available on all of these systems. This was the basis for a broad acceptance at industrial and commercial customers.

The increasing market share of industrial and commercial installations had several very critical implications for the HPC market. The manufacturers of supercomputers for numerical applications face in the market for small to medium size HPC systems the strong competition of manufacturers selling their systems in the very lucrative commercial market. These systems tend to have better price/performance ratios due to the larger production numbers of systems accepted at commercial customers and the reduced design costs of medium size systems. The market for the very high end systems itself is relatively small and does not grow strongly if at all. It cannot easily support specialized niche market manufacturers. This forces the remaining manufacturers to change the design for the very high end away from homogeneous large scale systems towards cluster concepts based on "off-the-shelf" components.

"Clusters" dominate as architecture in the TOP500. Some years ago in November 1999 we had only 7 clusters in the TOP500 while in June 2008 the list shows 400 cluster systems. At the same time the debate if we need new architectures for very high end supercomputers has increased in intensity again.

Novel hybrid architectures are likely to appear in the TOP500 list. The number one machine today, the IBM Roadrunner, is just such a system. The Roadrunner is a hybrid design built from commodity parts. The system is composed of two processor chip architectures, the IBM PowerXCell and the AMD

Opteron which use Infiniband interconnect. The system can be characterized as an Opteron based cluster with Cell accelerators. Each Opteron core has a Cell chip (composed of 9 cores). The Cell chip has 8 vector cores and a conventional PowerPC core. The vector cores provide the bulk of the computational performance.

4.1   Dynamic of the market

The HPC market is by its very nature very dynamic. This is not only reflected by the coming and going of new manufacturers but especially by the need to update and replace systems quite often to keep pace with the general performance increase. This general dynamic of the HPC market is well reflected in the TOP500. In Fig. 4 we show the number of systems, which fall off the end of the list within 6 month due to the increase in the entry level performance. We see an average replacement rate of about 180 systems every half year or more than half the list every year. This means that a system which is at position 100 at a given time will fall off the TOP500 within 2-3 years. The June 2008 list shows even a greater replacement with 301 systems being displaced from the previous list.

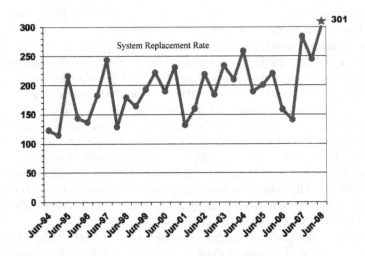

Fig. 4 The replacement rate in TOP500 defined as number of systems omitted because of their performance being too small.

4.2   Consumer and producer

The dynamic of the HPC market is well reflected in the rapidly changing market shares of the chip or system technologies, of manufacturers, customer types or application areas. If we however are interested in where these HPC systems are installed or produced we see a different picture.

Plotting the number of systems installed in different geographical areas in Fig. 5 we see a more or less steady distribution. The number of systems installed in the US is about half of the list, while the number of systems in Japan is slowly decreasing. Europe has again started to acquire HPC systems as shown in Fig. 5. While this can be interpreted as a reflection of increasing economical stamina of these countries it also highlights the fact that it is becoming easier for such countries to buy or even built cluster based systems themselves.

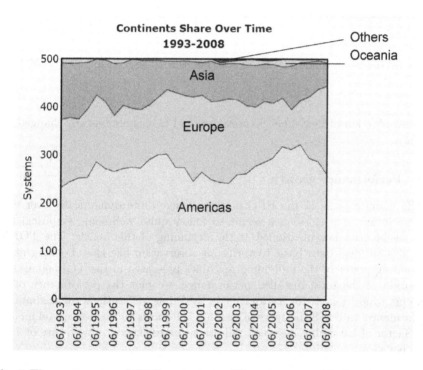

Fig. 5 The consumers of HPC systems in different geographical regions as seen in the TOP500.

Fig. 6 shows the decrease in the number of HPC systems in Japan and an initial use of such systems in China and India.

Looking at the producers of HPC system in Fig. 7 we see an even greater dominance of the US, which actually slowly increases over time. European manufacturers do not play any substantial role in the HPC market at all. Even

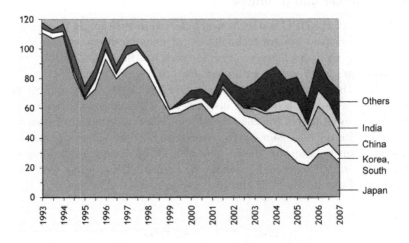

Fig. 6 The consumers of HPC systems in Asia as seen in the TOP500.

the introduction of new architectures such as PC clusters has not changed this picture.

4.3 Performance growth

While many aspects of the HPC market change quite dynamically over time, the evolution of performance seems to follow quite well some empirical laws such as Moore's law mentioned at the beginning of this article. The TOP500 provides an ideal data basis to verify an observation like this. Looking at the computing power of the individual machines presented in the TOP500 and the evolution of the total installed performance, we plot the performance of the systems at positions 1, 10, 100 and 500 in the list as well as the total accumulated performance. In Fig. 8 the curve of position 500 shows on the average an increase of a factor of 1.9 within one year. All other curves show a growth rate of 1.8 ± 0.05 per year.

4.4 Projections

Based on the current TOP500 data which cover the last thirteen years and the assumption that the current performance development continue for some time to come we can now extrapolate the observed performance and compare these values with the goals of the mentioned government programs. In Fig. 9 we extrapolate the observed performance values using linear regression on the logarithmic scale. This means that we fit exponential growth to all levels of performance in the TOP500. These simple fitting of the data shows surprisingly

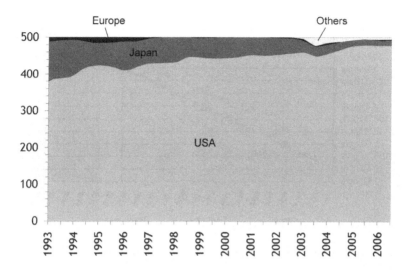

Fig. 7 The producers of HPC systems as seen in the TOP500.

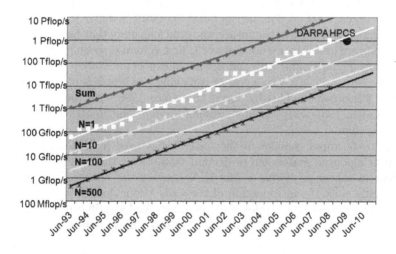

Fig. 8 Overall growth of accumulated and individual performance as seen in the TOP500.

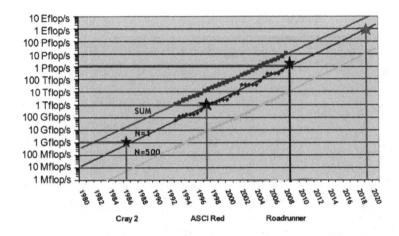

Fig. 9 Extrapolation of recent growth rates of performance as seen in the TOP500.

consistent results. In 1999 based on a similar extrapolation [6] we expected to have the first 100 TFlop/s system by 2005. We also predicted that by 2005 also no system smaller than 1 TFlop/s should be able to make the TOP500 any longer. Both of these predictions are basically certain to be fulfilled next year. Looking out another five years to 2010 we expected to see the first PetaFlops system at about 2009 [6]. We hit the PetaFlop mark in 2008.

Looking even further in the future we could speculate that based on the current doubling of performance every year the first system exceeding 100 Petaflop/s should be available around 2015 and we should expect an Exaflop system in 2019. Indeed we see an eleven year cycle of achieving three orders of magnitude increase in performance. This has been true since 1986 with the first Gigaflop system and in 1997 with the first Teraflop system and in 2008 with the first Petaflop system.

Due to the rapid changes in the technologies used in HPC systems there is however again no reasonable projection possible for the architecture of such a system in ten years. Even as the HPC market has changed its face quite substantially since the introduction of the Cray 1 four decades ago, there is no end in sight for these rapid cycles of re-definition. And we still can say that in the High Performance Computing Market "The Only Thing Constant Is Change".

References

1. G. Bell, The Next Ten Years of Supercomputing, *Proceedings 14th Supercomputer Conference*, Mannheim, June 10 -12, 1999, Editor: Hans Werner Meuer, CD-ROM (MaxionMedia), ISBN 3-932178-08-4.

2. R. W. Hockney, C. Jesshope, *Parallel Computers II: Architecture, Programming and Algorithms*, Adam Hilger, Ltd., Bristol, United Kingdom, 1988

3. H. W. Meuer, The Mannheim Supercomputer Statistics 1986–1992, *TOP500 Report 1993*, (University of Mannheim, 1994), 1–15.

4. H. W. Meuer, E. Strohmaier, J. J. Dongarra, and H. D. Simon, TOP500, www.top500.org.

5. H. D. Simon, High Performance Computing in the U.S., *TOP500 Report 1993*, (University of Mannheim, 1994), 116–147

6. E. Strohmaier, J.J. Dongarra, H.W. Meuer, and H.D. Simon, The marketplace of high-performance computing, *Parallel Computing* **25** no. 13 (1999) 1517–1544.

7. A. J. van der Steen, Overview of Recent Supercomputers, www.euroben.nl/-reports/overview07.pdf.

8. G. V. Wilson, Chronology of major developments in parallel computing and supercomputing, www.unipaderborn.de/fachbereich/AG/agmadh/WWW/GI/-History/history.txt.gz

9. P. R. Woodward, Perspectives on Supercomputing, *Computer*, **9** no. 10, (1996) 99–111

10. see: www.netlib.org/benchmark/performance.ps

Nonnegativity constraints in numerical analysis*

Donghui Chen[1] and Robert J. Plemmons[2]

[1] Department of Mathematics, Wake Forest University, Winston-Salem, NC 27109.
Presently at Dept. Mathematics Tufts University.
[2] Departments of Computer Science and Mathematics, Wake Forest University,
Winston-Salem, NC 27109.
Medford, MA 02155
plemmons@wfu.edu

Abstract. A survey of the development of algorithms for enforcing nonnegativity constraints in scientific computation is given. Special emphasis is placed on such constraints in least squares computations in numerical linear algebra and in nonlinear optimization. Techniques involving nonnegative low-rank matrix and tensor factorizations are also emphasized. Details are provided for some important classical and modern applications in science and engineering. For completeness, this report also includes an effort toward a literature survey of the various algorithms and applications of nonnegativity constraints in numerical analysis.

Key Words: nonnegativity constraints, nonnegative least squares, matrix and tensor factorizations, image processing, optimization.

1 Historical comments on enforcing nonnegativity

Nonnegativity constraints on solutions, or approximate solutions, to numerical problems are pervasive throughout science, engineering and business. In order to preserve inherent characteristics of solutions corresponding to amounts and measurements, associated with, for instance frequency counts, pixel intensities and chemical concentrations, it makes sense to respect the nonnegativity so as to avoid physically absurd and unpredictable results. This viewpoint has both computational as well as philosophical underpinnings. For example, for the sake of interpretation one might prefer to determine solutions from the same space, or a subspace thereof, as that of the input data.

In numerical linear algebra, nonnegativity constraints very often arise in least squares problems, which we denote as **nonnegative least squares** (NNLS). The design and implementation of NNLS algorithms has been the subject of considerable work the seminal book of Lawson and Hanson [45]. This book seems to contain the first widely used method for solving NNLS. A variation of their algorithm is available as **lsqnonneg** in Matlab. (For a history of NNLS computations in Matlab see [75].)

* Research supported by the Air Force Office of Scientific Research under grant FA9550-08-1-0151.

More recently, beginning in the 1990s, NNLS computations have been generalized to approximate nonnegative matrix or tensor factorizations, in order to obtain low-dimensional representations of nonnegative data. A suitable representation for data is essential to applications in fields such as statistics, signal and image processing, machine learning, and data mining. (See, e.g., the survey by Berry et al. [8].) Low rank constraints on high dimensional massive data sets are prevalent in dimensionality reduction and data analysis across numerous scientific disciplines. Techniques for dimensionality reduction and feature extraction include Principal Component Analysis (PCA), Independent Component Analysis (ICA), and (approximate) **Nonnegative Matrix Factorization** (NMF).

In this paper we are concerned primarily with NNLS as well as NMF and its extension to **Nonnegative Tensor Factorization** (NTF). A tensor can be thought of as a multi-way array, and our interest is in the natural extension of concepts involving data sets represented by 2-D arrays to 3-D arrays represented by tensors. Tensor analysis became immensely popular after Einstein used tensors as the natural language to describe laws of physics in a way that does not depend on the initial frame of reference. Recently, tensor analysis techniques have become a widely applied tool, especially in the processing of massive data sets. (See the work of Cichocki et al. [17] and Ho [32], as well as the program for the 2008 Stanford Workshop on Modern Massive Data Sets on the web page http://www.stanford.edu/group/mmds/.) Together, NNLS, NMF and NTF are used in various applications which will be discussed and referenced in this survey.

2   Preliminaries

We begin this survey with a review of some notation and terminology, some useful theoretical issues associated with nonnegative matrices arising in the mathematical sciences, and the Karush-Kuhn-Tucker conditions used in optimization. All matrices discussed are over the real numbers. For $A = (a_{ij})$ we write $A \geq 0$ if $a_{ij} \geq 0$ for each i and j. We say that A is a **nonnegative matrix**. The notation naturally extends to vectors, and to the term positive matrix.

Aspects of the theory of nonnegative matrices, such as the classical Perron-Frobenius theory, have been included in various books. For more details the reader is referred to the books, in chronological order, by Varga [83], by Berman and Plemmons. [7], and by Bapat and Raghavan [1]. This topic leads naturally to the concepts of inverse-positivity, monotonicity and iterative methods, and M-matrix computations. For example, M-Matrices A have positive diagonal entries and non-positive off-diagonal entries, with the added condition that A^{-1} is a nonnegative matrix. Associated linear systems of equations $Ax = b$ thus have nonnegative solutions whenever $b \geq 0$. Applications of M-Matrices abound in numerical analysis topics such as numerical PDEs and Markov Chain analysis, as well as in economics, operations research, and statistics, see e.g., [7,83].

For the sake of completeness we state the classical Perron-Frobenious Theorem for irreducible nonnegative matrices. Here, an $n \times n$ matrix A is said to be

reducible if $n \geq 2$ and there exists a permutation matrix $\mathbf{P}$ such that

$$\mathbf{PAP}^T = \begin{bmatrix} \mathbf{B} & 0 \\ \mathbf{C} & \mathbf{D} \end{bmatrix}, \tag{2.1}$$

where $\mathbf{B}$ and $\mathbf{D}$ are square matrices and 0 is a zero matrix. The matrix $\mathbf{A}$ is **irreducible** if it is not reducible.

Perron-Frobenius theorem:

Let A be a $n \times n$ nonnegative irreducible matrix. Then there exists a real number $\lambda_0 > 0$ and a positive vector y such that

- $\mathbf{Ay} = \lambda_0 \mathbf{y}$.
- The eigenvalue λ_0 is geometrically simple. That is, any two eigenvectors corresponding to λ_0 are linearly dependent.
- The eigenvalue λ_0 is maximal in modulus among all the eigenvalues of $\mathbf{A}$. That is, for any eigenvalue μ of $\mathbf{A}$, $|\mu| \leq \lambda_0$.
- The only nonnegative, nonzero eigenvectors of $\mathbf{A}$ are just the positive scalar multiplies of y.
- The eigenvalue λ_0 is algebraically simple. That is, λ_0 is a simple root of the characteristic polynomial of $\mathbf{A}$.
- Let $\lambda_0, \lambda_1, \ldots, \lambda_{k-1}$ be the distinct eigenvalues of $\mathbf{A}$ with $|\lambda_i| = \lambda_0$, $i = 1, 2, \ldots, k-1$. Then they are precisely the solutions of the equation $\lambda^k - \lambda_0^k = 0$.

As a simple illustration of one application of this theorem, we mention that a finite irreducible Markov process associated with a probability matrix $\mathbf{S}$ must have a positive stationary distribution vector, which is associated with the eigenvalue 1 of $\mathbf{S}$. (See, e.g., [7].)

Another concept that will be useful in this paper is the classical Karush-Kuhn-Tucker conditions (also known as the Kuhn-Tucker or the KKT conditions). The set of conditions is a generalization of the method of Lagrange multipliers.

Karush-Kuhn-Tucker conditions:

The Karush-Kuhn-Tucker (KKT) conditions are necessary for a solution in nonlinear programming to be optimal. Consider the following nonlinear optimization problem:

Let $\mathbf{x}^*$ be a local minimum of

$$\min_{\mathbf{x}} f(\mathbf{x}) \text{ subject to } \begin{cases} h(\mathbf{x}) = 0 \\ g(\mathbf{x}) \leq 0 \end{cases}$$

and suppose $\mathbf{x}^*$ is a regular point for the constraints, i.e. the Jacobian of the binding constraints at that point is of full rank. Then $\exists \ \lambda$ and μ such that

$$\nabla f(\mathbf{x}^*) + \lambda^T \nabla h(\mathbf{x}^*) + \mu^T \nabla g(\mathbf{x}^*) = 0$$

$$\mu^T g(\mathbf{x}^*) = 0$$

$$h(\mathbf{x}^*) = 0 \qquad\qquad (2.2)$$

$$\mu \geq 0.$$

Next we move to the topic of least squares computations with nonnegativity constraints, NNLS. Both old and new algorithms are outlined. We will see that NNLS leads in a natural way to the topics of approximate low-rank nonnegative matrix and tensor factorizations, NMF and NTF.

3   Nonnegative least squares

3.1   Introduction

A fundamental problem in data modeling is the estimation of a parameterized model for describing the data. For example, imagine that several experimental observations that are linear functions of the underlying parameters have been made. Given a sufficiently large number of such observations, one can reliably estimate the true underlying parameters. Let the unknown model parameters be denoted by the vector $\mathbf{x} = (x_1, \cdots, x_n)^T$, the different experiments relating $\mathbf{x}$ be encoded by the measurement matrix $\mathbf{A} \in R^{m \times n}$, and the set of observed values be given by $\mathbf{b}$. The aim is to reconstruct a vector $\mathbf{x}$ that explains the observed values as well as possible. This requirement may be fulfilled by considering the linear system

$$\mathbf{A}\mathbf{x} = \mathbf{b},$$

where the system may be either under-determined ($m < n$) or over-determined ($m \geq n$). In the latter case, the technique of least-squares proposes to compute $\mathbf{x}$ so that the reconstruction error

$$f(\mathbf{x}) = \frac{1}{2} \|\mathbf{A}\mathbf{x} - \mathbf{b}\|^2 \qquad\qquad (3.1)$$

is minimized, where $\| \cdot \|$ denotes the $\mathbf{L}_2$ norm. However, the estimation is not always that straightforward because in many real-world problems the underlying parameters represent quantities that can take on only nonnegative values, e.g., amounts of materials, chemical concentrations, pixel intensities, to name a few. In such a case, problem (3.1) must be modified to include nonnegativity constraints on the model parameters $\mathbf{x}$. The resulting problem is called Nonnegative Least Squares (NNLS), and is formulated as follows:

NNLS problem:

Given a matrix $\mathbf{A} \in R^{m \times n}$ *and the set of observed values given by* $\mathbf{b} \in R^m$, *find a nonnegative a vector* $\mathbf{x} \in R^n$ *to minimize the functional* $f(\mathbf{x}) = \frac{1}{2}\|\mathbf{Ax} - \mathbf{b}\|^2$, *i.e.*

$$\min_{\mathbf{x}} f(\mathbf{x}) = \frac{1}{2}\|\mathbf{Ax} - \mathbf{b}\|^2, \tag{3.2}$$
$$subject \ \ to \ \ \mathbf{x} \geq 0.$$

The gradient of $f(\mathbf{x})$ is $\nabla f(\mathbf{x}) = \mathbf{A}^T(\mathbf{Ax} - \mathbf{b})$ and the KKT optimality conditions for NNLS problem (3.2) are

$$\mathbf{x} \geq 0$$
$$\nabla f(\mathbf{x}) \geq 0 \tag{3.3}$$
$$\nabla f(\mathbf{x})^T \mathbf{x} = 0.$$

Some of the iterative methods for solving (3.2) are based on the solution of the corresponding linear complementarity problem (LCP).

Linear Complementarity Problem:

Given a matrix $\mathbf{A} \in R^{m \times n}$ *and the set of observed values be given by* $\mathbf{b} \in R^m$, *find a vector* $\mathbf{x} \in R^n$ *to minimize the functional*

$$\lambda = \nabla f(\mathbf{x}) = \mathbf{A}^T \mathbf{Ax} - \mathbf{A}^T \mathbf{b} \geq 0$$
$$\mathbf{x} \geq 0 \tag{3.4}$$
$$\lambda^T \mathbf{x} = 0.$$

Problem (3.4) is essentially the set of KKT optimality conditions (3.3) for quadratic programming. The problem reduces to finding a nonnegative $\mathbf{x}$ which satisfies $(\mathbf{Ax} - \mathbf{b})^T \mathbf{Ax} = 0$. Handling nonnegative constraints is computationally nontrivial because we are dealing with expansive nonlinear equations. An equivalent but sometimes more tractable formulation of NNLS using the residual vector variable $\mathbf{p} = \mathbf{b} - \mathbf{Ax}$ is as follows:

$$\min_{\mathbf{x},\mathbf{p}} \frac{1}{2}\mathbf{p}^T \mathbf{p} \tag{3.5}$$
$$s. \ t. \ \mathbf{Ax} + \mathbf{p} = \mathbf{b}, \ \ \mathbf{x} \geq 0.$$

The advantage of this formulation is that we have a simple and separable objective function with linear and nonnegativity constraints.

The NNLS problem is fairly old. The algorithm of Lawson and Hanson [45] seems to be the first method to solve it. (This algorithm is available as the **lsqnonneg** in Matlab, see [75].) An interesting thing about NNLS is that it is solved iteratively, but as Lawson and Hanson show, the iteration always converges and terminates. There is no cutoff in iteration required. Sometimes it might run too long, and have to be terminated, but the solution will still be "fairly good", since the solution improves smoothly with iteration. Noise, as expected, increases the number of iterations required to reach the solution.

3.2   Numerical approaches and algorithms

Over the years a variety of methods have been applied to tackle the NNLS problem. Although those algorithms can straddle more than one class, in general they can be roughly divided into active-set methods and iterative approaches. (See Table 1 for a listing of some approaches to solving the NNLS problem.)

Table 1 Some Numerical Approaches and Algorithms for NNLS

Active Set Methods	Iterative Approaches	Other Methods
lsqnonneg in Matlab	Projected Quasi-Newton NNLS	Interior Point Method
Bro and de Jong's Fast NNLS	Projected Landweber method	Principal Block Pivoting method
Fast Combinatorial NNLS	Sequential Coordinate-wise Alg.	

3.2.1   Active-set methods

Active-set methods [24] are based on the observation that only a small subset of constraints are usually active (i.e. satisfied exactly) at the solution. There are n inequality constraints in NNLS problem. The ith constraint is said to be *active*, if the ith regression coefficient will be is negative (or zero) if unconstrained, otherwise the constraint is passive. An *active set algorithm* uses the fact that if the true active set is known, the solution to the least squares problem will simply be the unconstrained least squares solution to the problem using only the variables corresponding to the passive set, setting the regression coefficients of the active set to zero. This can also be stated as: if the active set is known, the solution to the NNLS problem is obtained by treating the active constraints as equality constraints rather than inequality constraints. To find this solution, an alternating least squares algorithm is applied. An initial feasible set of regression coefficients is found. A feasible vector is a vector with no elements violating the constraints. In this case the vector containing only zeros is a feasible starting vector as it contains no negative values. In each step of the algorithm, variables are identified and removed from the active set in such a way that the least least squares fit strictly decreases. After a finite number of iterations the true active set is found and the solution is found by simple linear regression on the unconstrained subset of the variables.

The NNLS algorithm of Lawson and Hanson [45] is an *active set method*, and was the *de facto* method for solving (3.2) for many years. Recently, Bro and de Jong [12] modified it and developed a method called Fast NNLS (FNNLS)), which often speeds up the basic algorithm, especially in the presence of multiple right-hand sides, by avoiding unnecessary re-computations. A recent variant of FNNLS, called fast combinatorial NNLS [80], appropriately rearranges calculations to achieve further speedups in the presence of multiple right hand sides. However, all of these approaches still depend on $\mathbf{A}^T\mathbf{A}$, or the normal equations in factored form, which is infeasible for ill-conditioned problems.

Lawson and Hanson's algorithm:

In their landmark text [45], Lawson and Hanson give the Standard algorithm for NNLS which is an *active set method* [24]. Mathworks [75] modified the algorithm NNLS, which ultimately was renamed to *"lsqnonneg"*.

Notation: The matrix $\mathbf{A}^P$ is a matrix associated with only the variables currently in the passive set P.

Algorithm *lsqnonneg* :

Input: $\mathbf{A} \in \mathbf{R}^{m \times n}$, $\mathbf{b} \in \mathbf{R}^m$
Output: $\mathbf{x}^* \geq 0$ such that $\mathbf{x}^* = \arg\min \|\mathbf{A}\mathbf{x} - \mathbf{b}\|^2$.
Initialization: $P = \varnothing$, $R = \{1, 2, \cdots, n\}$, $\mathbf{x} = \mathbf{0}$, $\mathbf{w} = \mathbf{A}^T(\mathbf{b} - \mathbf{A}\mathbf{x})$
repeat

1. Proceed if $R \neq \varnothing \wedge [\max_{i \in R}(w_i) > tolerance]$
2. $j = \arg\max_{i \in R}(w_i)$
3. Include the index j in P and remove it from R
4. $\mathbf{s}^P = [(\mathbf{A}^P)^T\mathbf{A}^P]^{-1}(\mathbf{A}^P)^T\mathbf{b}$
   4.1. Proceed if $\min(\mathbf{s}^P) \leq 0$
   4.2. $\alpha = -\min_{i \in P}[x_i/(x_i - s_i)]$
   4.3. $\mathbf{x} := \mathbf{x} + \alpha(\mathbf{s} - \mathbf{x})$
   4.4. Update R and P
   4.5. $\mathbf{s}^P = [(\mathbf{A}^P)^T\mathbf{A}^P]^{-1}(\mathbf{A}^P)^T\mathbf{b}$
   4.6. $\mathbf{s}^R = \mathbf{0}$
5. $\mathbf{x} = \mathbf{s}$
6. $\mathbf{w} = \mathbf{A}^T(\mathbf{b} - \mathbf{A}\mathbf{x})$

It is proved by Lawson and Hanson that the iteration of the NNLS algorithm is finite. Given sufficient time, the algorithm will reach a point where the Kuhn-Tucker conditions are satisfied, and it will terminate. There is no arbitrary cutoff in iteration required; in that sense it is a direct algorithm. It is not direct in the sense that the upper limit on the possible number of iterations that the algorithm might need to reach the point of optimum solution is impossibly large. There is no good way of telling exactly how many iterations it will require in a practical sense. The solution does improve smoothly as the iteration continues. If it is terminated early, one will obtain a sub-optimal but likely still fairly good image.

However, when applied in a straightforward manner to large scale NNLS problems, this algorithm's performance is found to be unacceptably slow owing to the need to perform the equivalent of a full pseudo-inverse calculation for each observation vector. More recently, Bro and de Jong [12] have made a substantial speed improvement to Lawson and Hanson's algorithm for the case of a large number of observation vectors, by developing a modified NNLS algorithm.

Fast NNLS *fnnls* :

In the paper [12], Bro and de Jong give a modification of the standard algorithm for NNLS by Lawson and Hanson. Their algorithm, called Fast Nonnegative Least Squares, *fnnls*, is specifically designed for use in multiway decomposition methods for tensor arrays such as PARAFAC and N-mode PCA (See the material on tensors given later in this paper.) They realized that large parts of the pseudo-inverse could be computed once but used repeatedly. Specifically, their algorithm precomputes the cross-product matrices that appear in the normal equation formulation of the least squares solution. They also observed that, during alternating least squares (ALS) procedures (to be discussed later), solutions tend to change only slightly from iteration to iteration. In an extension to their NNLS algorithm that they characterized as being for "advanced users", they retained information about the previous iteration's solution and were able to extract further performance improvements in ALS applications that employ NNLS. These innovations led to a substantial performance improvement when analyzing large multivariate, multiway data sets.

Algorithm *fnnls* :

Input: $\mathbf{A} \in \mathbf{R}^{m \times n}$, $\mathbf{b} \in \mathbf{R}^m$
Output: $\mathbf{x}^* \geq 0$ such that $\mathbf{x}^* = \arg \min \|\mathbf{A}\mathbf{x} - \mathbf{b}\|^2$.
Initialization: $P = \varnothing, R = \{1, 2, \cdots, n\}, \mathbf{x} = \mathbf{0}, \mathbf{w} = \mathbf{A}^T \mathbf{b} - (\mathbf{A}^T \mathbf{A})\mathbf{x}$
repeat

1. Proceed if $R \neq \varnothing \wedge [\max_{i \in R}(w_i) > tolerance]$
2. $j = \arg \max_{i \in R}(w_i)$
3. Include the index j in P and remove it from R
4. $\mathbf{s}^P = [(\mathbf{A}^T \mathbf{A})^P]^{-1}(\mathbf{A}^T \mathbf{b})^P$
    4.1. Proceed if $\min(\mathbf{s}^P) \leq 0$
    4.2. $\alpha = -\min_{i \in P}[x_i/(x_i - s_i)]$
    4.3. $\mathbf{x} := \mathbf{x} + \alpha(\mathbf{s} - \mathbf{x})$
    4.4. Update R and P
    4.5. $\mathbf{s}^P = [(\mathbf{A}^T \mathbf{A})^P]^{-1}(\mathbf{A}^T \mathbf{b})^P$
    4.6. $\mathbf{s}^R = \mathbf{0}$
5. $\mathbf{x} = \mathbf{s}$
6. $\mathbf{w} = \mathbf{A}^T(\mathbf{b} - \mathbf{A}\mathbf{x})$

While Bro and de Jong's algorithm precomputes parts of the pseudo-inverse, the algorithm still requires work to complete the pseudo-inverse calculation once for each vector observation. A recent variant of *fnnls*, called fast combinatorial NNLS [80], appropriately rearranges calculations to achieve further speedups in the presence of multiple observation vectors $\mathbf{b}_i, i = 1, 2, \ldots, l$. This new method rigorously solves the constrained least squares problem while exacting essentially no performance penalty as compared with Bro and de Jong's algorithm. The new algorithm employs combinatorial reasoning to identify and group together all observations $\mathbf{b}_i$ that share a common pseudo-inverse at each stage in the NNLS iteration. The complete pseudo-inverse is then computed just once per group and, subsequently, is applied individually to each observation in the group. As a result, the computational burden is significantly reduced and the time required to perform ALS operations is likewise reduced. Essentially, if there is only one observation, this new algorithm is no different from Bro and de Jong's algorithm.

In the paper [19], Dax concentrates on two problems that arise in the implementation of an active set method. One problem is the choice of a good starting point. The second problem is how to move away from a *"dead point"*. The results of his experiments indicate that the use of *Gauss-Seidel* iterations to obtain a starting point is likely to provide large gains in efficiency. And also, dropping one constraint at a time is advantageous to dropping several constraints at a time.

However, all these *active set methods* still depend on the normal equations, rendering them infeasible for ill-conditioned. In contrast to an active set method, iterative methods, for instance gradient projection, enables one to incorporate multiple active constraints at each iteration.

3.2.2   Algorithms based on iterative methods

The main advantage of this class of algorithms is that by using information from a projected gradient step along with a good guess of the active set, one can handle multiple active constraints per iteration. In contrast, the active-set method typically deals with only one active constraint at each iteration. Some of the iterative methods are based on the solution of the corresponding LCP (3.4). In contrast to an active set approach, iterative methods like gradient projection enables the incorporation of multiple active constraints at each iteration.

Projective quasi-Newton NNLS (PQN-NNLS)

In the paper [41], Kim, et al. proposed a projection method with non-diagonal gradient scaling to solve the NNLS problem (3.2). In contrast to an active set approach, gradient projection avoids the pre-computation of $\mathbf{A}^T\mathbf{A}$ and $\mathbf{A}^T\mathbf{b}$, which is required for the use of the active set method *fnnls*. It also enables their method to incorporate multiple active constraints at each iteration. By employing non-diagonal gradient scaling, **PQN-NNLS** overcomes some of the deficiencies of a projected gradient method such as slow convergence and zigzagging. An important characteristic of **PQN-NNLS** algorithm is that despite the efficiencies, it still remains relatively simple in comparison with other optimization-oriented algorithms. Also in this paper, Kim et al. gave experiments to show that their method outperforms other standard approaches to solving the NNLS problem, especially for large-scale problems.

Algorithm PQN-NNLS:

Input: $\mathbf{A} \in \mathbf{R}^{m \times n}$, $\mathbf{b} \in \mathbf{R}^m$
Output: $\mathbf{x}^* \geq 0$ such that $\mathbf{x}^* = \arg\min \|\mathbf{A}\mathbf{x} - \mathbf{b}\|^2$.
Initialization: $\mathbf{x}^0 \in \mathbf{R}^n_+$, $\mathbf{S}^0 \leftarrow \mathbf{I}$ and $k \leftarrow 0$
repeat

1. Compute fixed variable set $\mathbf{I}^k = \{i : x_i^k = 0, [\nabla f(\mathbf{x}^k)]_i > 0\}$
2. Partition $\mathbf{x}^k = [\mathbf{y}^k; \mathbf{z}^k]$, where $y_i^k \notin \mathbf{I}^k$ and $z_i^k \in \mathbf{I}^k$
3. Solve equality-constrained subproblem:
    3.1. Find appropriate values for α^k and β^k
    3.2. $\gamma^k(\beta^k; \mathbf{y}^k) \leftarrow \mathcal{P}(\mathbf{y}^k - \beta^k \bar{\mathbf{S}}^k \nabla f(\mathbf{y}^k))$
    3.3. $\tilde{\mathbf{y}} \leftarrow \mathbf{y}^k + \alpha(\gamma^k(\beta^k; \mathbf{y}^k) - \mathbf{y}^k)$
4. Update gradient scaling matrix $\mathbf{S}^k$ to obtain $\mathbf{S}^{k+1}$
5. Update $\mathbf{x}^{k+1} \leftarrow [\tilde{\mathbf{y}}; \mathbf{z}^k]$
6. $k \leftarrow k + 1$

until Stopping criteria are met.

Sequential coordinate-wise algorithm for NNLS

In [23], the authors propose a novel sequential coordinate-wise (SCA) algorithm which is easy to implement and it is able to cope with large scale problems. They also derive stopping conditions which allow control of the distance of the solution found to the optimal one in terms of the optimized objective function. The algorithm produces a sequence of vectors $\mathbf{x}^0, \mathbf{x}^1, \ldots, \mathbf{x}^t$ which converges to the optimal x^*. The idea is to optimize in each iteration with respect to a single coordinate while the remaining coordinates are fixed. The optimization with respect to a single coordinate has an analytical solution, thus it can be computed efficiently.

Notation: $\mathcal{I} = \{1, 2, \cdots, n\}$, $\mathcal{I}_k = \mathcal{I}/k$, $\mathbf{H} = \mathbf{A}^T \mathbf{A}$ which is semi-positive definite, and $\mathbf{h}_k$ denotes the kth column of $\mathbf{H}$.

Algorithm SCA-NNLS:

Input: $\mathbf{A} \in \mathbf{R}^{m \times n}$, $\mathbf{b} \in \mathbf{R}^m$
Output: $\mathbf{x}^* \geq 0$ such that $\mathbf{x}^* = \arg \min \|\mathbf{Ax} - \mathbf{b}\|^2$.
Initialization: $\mathbf{x}^0 = \mathbf{0}$ and $\mu^0 = f = -\mathbf{A}^T \mathbf{b}$
repeat For $k = 1$ to n

1. $\mathbf{x}_k^{t+1} = \max\left(0, \mathbf{x}_k^t - \frac{\mu_k^t}{\mathbf{H}_{k,k}}\right)$, and $\mathbf{x}_i^{t+1} = \mathbf{x}_i^t$, $\forall i \in \mathcal{I}_k$
2. $\mu^{t+1} = \mu^t + (\mathbf{x}_k^{t+1} - \mathbf{x}_k^t)\mathbf{h}_k$

until Stopping criteria are met.

3.2.3   Other methods:

Principal block pivoting method

In the paper [13], the authors gave a block principal pivoting algorithm for large and sparse NNLS. They considered the linear complementarity problem (3.4). The n indices of the variables in $\mathbf{x}$ are divided into complementary sets F and G, and let $\mathbf{x}_F$ and $\mathbf{y}_G$ denote pairs of vectors with the indices of their nonzero entries in these sets. Then the pair $(\mathbf{x}_F, \mathbf{y}_G)$ is a *complementary basic solution* of Equation (3.4) if $\mathbf{x}_F$ is a solution of the unconstrained least squares problem

$$\min_{\mathbf{x}_F \in \mathbb{R}^{|F|}} \|\mathbf{A}_F \mathbf{x}_F - \mathbf{b}\|_2^2 \tag{3.6}$$

where $\mathbf{A}_F$ is formed from $\mathbf{A}$ by selecting the columns indexed by F, and $\mathbf{y}_G$ is obtained by

$$\mathbf{y}_G = \mathbf{A}_G^T(\mathbf{A}_F \mathbf{x}_F - \mathbf{b}). \tag{3.7}$$

If $\mathbf{x}_F \geq 0$ and $\mathbf{y}_G \geq 0$, then the solution is *feasible*. Otherwise it is *infeasible*, and we refer to the negative entries of $\mathbf{x}_F$ and $\mathbf{y}_G$ as *infeasible variables*. The idea of the algorithm is to proceed through infeasible complementary basic solutions of (3.4) to the unique feasible solution by exchanging infeasible variables between F and G and updating $\mathbf{x}_F$ and $\mathbf{y}_G$ by (3.6) and (3.7). To minimize the number of solutions of the least-squares problem in (3.6), it is desirable to exchange variables in large groups if possible. The performance of the algorithm is several times faster than Matstoms' Matlab implementation [51] of the same algorithm. Further, it matches the accuracy of Matlab's built-in *lsqnonneg* function. (The program is available online at http://plato.asu.edu/sub/nonlsq.html).

Block principal pivoting algorithm:

Input: $A \in \mathbf{R}^{m \times n}$, $b \in \mathbf{R}^m$
Output: $x^* \geq 0$ such that $x^* = \arg\min \|Ax - b\|^2$.
Initialization: $F = \varnothing$ and $G = 1, \ldots, n, x = 0, y = -A^T b$, and $p = 3, N = \infty$.
repeat:

1. Proceed if (x_F, y_G) is an infeasible solution.
2. Set n to the number of negative entries in x_F and y_G.
   2.1 Proceed if $n < N$,
      2.1.1 Set $N = n, p = 3$,
      2.1.2 Exchange all infeasible variables between F and G.
   2.2 Proceed if $n \geq N$
      2.2.1 Proceed if $p > 0$,
         2.2.1.1 set $p = p - 1$
         2.2.1.2 Exchange all infeasible variables between F and G.
      2.2.2 Proceed if $p \leq 0$,
         2.2.2.1 Exchange only the infeasible variable with largest index.
3. Update x_F and y_G by Equation (3.6) and (3.7).
4. Set Variables in $x_F < 10^{-12}$ and $y_G < 10^{-12}$ to zero.

Interior point Newton-like method:

In addition to the methods above, Interior Point methods can be used to solve NNLS problems. They generate an infinite sequence of strictly feasible points converging to the solution and are known to be competitive with active set methods for medium and large problems. In the paper [4], the authors present an interior-point approach suited for NNLS problems. Global and locally fast convergence is guaranteed even if a degenerate solution is approached and the structure of the given problem is exploited both in the linear algebra phase and in the globalization strategy. Viable approaches for implementation are discussed and numerical results are provided. Here we give an interior algorithm for NNLS, more detailed discussion could be found in the paper [4].

Notation: $g(x)$ is the gradient of the objective function (3.2), i.e. $g(x) = \nabla f(x) = A^T(Ax - b)$. Therefore, by the KKT conditions, x^* can be found by searching for the positive solution of the system of nonlinear equations

$$D(x)g(x) = 0, \tag{3.8}$$

where $D(x) = \mathrm{diag}(d_1(x), \ldots, d_n(x))$, has entries

$$d_i(x) = \begin{cases} x_i & \text{if } g_i(x) \geq 0, \\ 1 & \text{otherwise.} \end{cases} \tag{3.9}$$

The matrix $W(x)$ is defined by $W(x) = \text{diag}(w_1(x), \ldots, w_n(x))$, where $w_i(x) = \frac{1}{d_i(x) + e_i(x)}$ and for $1 < s \leq 2$

$$e_i(x) = \begin{cases} g_i(x) & \text{if } 0 \leq g_i(x) < x_i^s \text{ or } g_i(x)^s > x_i, \\ 1 & \text{otherwise}. \end{cases} \tag{3.10}$$

Newton Like method for NNLS:

Input: $\mathbf{A} \in \mathbf{R}^{m \times n}$, $\mathbf{b} \in \mathbf{R}^m$
Output: $\mathbf{x}^* \geq 0$ such that $\mathbf{x}^* = \arg\min \|\mathbf{A}\mathbf{x} - \mathbf{b}\|^2$.
Initialization: $\mathbf{x}_0 > \mathbf{0}$ and $\sigma < 1$
repeat

1. Choose $\eta_k \in [0, 1)$
2. Solve $Z_k \tilde{p} = -W_k^{\frac{1}{2}} D_k^{\frac{1}{2}} g_k + \tilde{r}_k$, $\|\tilde{r}_k\|_2 \leq \eta_k \|W_k D_k g_k\|_2$
3. Set $p = W_k^{\frac{1}{2}} D_k^{\frac{1}{2}} \tilde{p}$
4. Set $p_k = max\{\sigma, 1 - \|p(x_k + p) - x_k\|_2\}(p(x_k + p) - x_k)$
5. Set $x_{k+1} = x_k + p_k$

until Stopping criteria are met.

We next move to the extension of Problem NNLS to approximate low-rank nonnegative matrix factorization and later extend that concept to approximate low-rank nonnegative tensor (multiway array) factorization.

4   Nonnegative matrix and tensor factorizations

As indicated earlier, NNLS leads in a natural way to the topics of approximate nonnegative matrix and tensor factorizations, NMF and NTF. We begin by discussing algorithms for approximating an $m \times n$ nonnegative matrix $\mathbf{X}$ by a low-rank matrix, say $\mathbf{Y}$, that is factored into $\mathbf{Y} = \mathbf{W}\mathbf{H}$, where $\mathbf{W}$ has $k \leq \min\{m, n\}$ columns, and $\mathbf{H}$ has k rows.

4.1   Nonnegative matrix factorization

In Nonnegative Matrix Factorization (NMF), an $m \times n$ (nonnegative) mixed data matrix $\mathbf{X}$ is approximately factored into a product of two nonnegative rank-k matrices, with k small compared to m and n, $\mathbf{X} \approx \mathbf{W}\mathbf{H}$. This factorization has the advantage that $\mathbf{W}$ and $\mathbf{H}$ can provide a physically realizable representation of the mixed data. NMF is widely used in a variety of applications, including air emission control, image and spectral data processing, text mining, chemometric analysis, neural learning processes, sound recognition, remote sensing, and object characterization, see, e.g. [8].

NMF problem: *Given a nonnegative matrix* $X \in \mathbf{R}^{m \times n}$ *and a positive integer* $k \leq \min\{m, n\}$, *find nonnegative matrices* $W \in \mathbf{R}^{m \times k}$ *and* $H \in \mathbf{R}^{k \times n}$ *to minimize the function* $f(\mathbf{W}, \mathbf{H}) = \frac{1}{2}\|X - \mathbf{WH}\|_F^2$, *i.e.*

$$\min_{\mathbf{H}} f(\mathbf{H}) = \|X - \sum_{i=1}^{k} \mathbf{W}^{(i)} \circ \mathbf{H}^{(i)}\| \quad subject \quad to \quad \mathbf{W}, \mathbf{H} \geq 0 \qquad (4.1)$$

where $'\circ'$ *denotes outer product,* $\mathbf{W}^{(i)}$ *is ith column of* $\mathbf{W}$, $\mathbf{H}^{(i)}$ *is ith column of* $\mathbf{H}^{\mathbf{T}}$

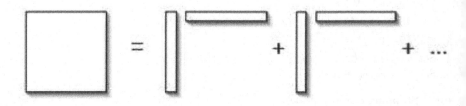

Fig. 1 An illustration of nonnegative matrix factorization.

See Figure 1 which provides an illustration of matrix approximation by a sum of rank one matrices determined by **W** and **H**. The sum is truncated after k terms.

Quite a few numerical algorithms have been developed for solving the NMF. The methodologies adapted are following more or less the principles of alternating direction iterations, the projected Newton, the reduced quadratic approximation, and the descent search. Specific implementations generally can be categorized into alternating least squares algorithms [57], multiplicative update algorithms [35, 46, 47], gradient descent algorithms, and hybrid algorithms [60, 62]. Some general assessments of these methods can be found in [15, 50]. It appears that there is much room for improvement of numerical methods. Although schemes and approaches are different, any numerical method is essentially centered around satisfying the first order optimality conditions derived from the Kuhn-Tucker theory. Note that the computed factors W and H may only be local minimizers of (4.1).

Theorem 4.1. *Necessary conditions for* $(W, H) \in \mathbf{R}_+^{m \times p} \times \mathbf{R}_+^{p \times n}$ *to solve the nonnegative matrix factorization problem (4.1) are*

$$
\begin{aligned}
W. * ((X - WH)H^T) &= \mathbf{0} \in \mathbf{R}^{m \times p}, \\
H. * (W^T(X - WH)) &= \mathbf{0} \in \mathbf{R}^{p \times n}, \\
(X - WH)H^T &\leq \mathbf{0}, \\
W^T(X - WH) &\leq \mathbf{0},
\end{aligned}
\tag{4.2}
$$

where '.' denotes the Hadamard product.*

Alternating Least Squares (ALS) algorithms for NMF

Since the Frobenius norm of a matrix is just the sum of Euclidean norms over columns (or rows), minimization or descent over either **W** or **H** boils down to solving a sequence of nonnegative least squares (NNLS) problems. In the class of ALS algorithms for NMF, a least squares step is followed by another least squares step in an alternating fashion, thus giving rise to the ALS name. ALS algorithms were first used by Paatero [57], exploiting the fact that, while the optimization problem of (4.1) is not convex in both **W** and **H**, it is convex in either **W** or **H**. Thus, given one matrix, the other matrix can be found with NNLS computations. An elementary ALS algorithm in matrix notation follows.

ALS algorithm for NMF:

Initialization: Let **W** be a random matrix $\mathbf{W} = rand(m, k)$ or use another initialization from [44]
repeat: for $i = 1 : maxiter$

1. (NNLS) Solve for **H** in the matrix equation $\mathbf{W}^T\mathbf{W}\mathbf{H} = \mathbf{W}^T\mathbf{X}$ by solving

$$
\min_{\mathbf{H}} f(\mathbf{H}) = \frac{1}{2}\|\mathbf{X} - \mathbf{W}\mathbf{H}\|_F^2 \quad subject \ to \ \mathbf{H} \geq 0,
$$

with **W** fixed,
2. (NNLS) Solve for **W** in the matrix equation $\mathbf{H}\mathbf{H}^T\mathbf{W}^T = \mathbf{H}\mathbf{X}^T$ by solving

$$
\min_{\mathbf{W}} f(\mathbf{W}) = \frac{1}{2}\|\mathbf{X}^T - \mathbf{H}^T\mathbf{W}^T\|_F^2 \quad subject \ to \ \mathbf{W} \geq 0
$$

with **H** fixed.

end

Compared to other methods for NMF, the ALS algorithms are more flexible, allowing the iterative process to escape from a poor path. Depending on the implementation, ALS algorithms can be very fast. The implementation shown above requires significantly less work than other NMF algorithms and slightly less work than an SVD implementation. Improvements to the basic ALS algorithm appear in [44, 58].

We conclude this section with a discussion of the convergence of ALS algorithms. Algorithms following an alternating process, approximating $\mathbf{W}$, then $\mathbf{H}$, and so on, are actually variants of a simple optimization technique that has been used for decades, and are known under various names such as alternating variables, coordinate search, or the method of local variation [55]. While statements about global convergence in the most general cases have not been proven for the method of alternating variables, a bit has been said about certain special cases. For instance, [64] proved that every limit point of a sequence of alternating variable iterates is a stationary point. Others [66, 67, 84] proved convergence for special classes of objective functions, such as convex quadratic functions. Furthermore, it is known that an ALS algorithm that properly enforces nonnegativity, for example, through the nonnegative least squares (NNLS) algorithm of Lawson and Hanson [45], will converge to a local minimum [10, 26, 49].

4.2   Nonnegative tensor decomposition

Nonnegative Tensor Factorization (NTF) is a natural extension of NMF to higher dimensional data. In NTF, high-dimensional data, such as hyperspectral or other image cubes, is factored directly, it is approximated by a sum of rank 1 nonnegative tensors. The ubiquitous tensor approach, originally suggested by Einstein to explain laws of physics without depending on inertial frames of reference, is now becoming the focus of extensive research. Here, we develop and apply NTF algorithms for the analysis of spectral and hyperspectral image data. The algorithm given here combines features from both NMF and NTF methods.

Notation: The symbol $*$ denotes the **Hadamard** (i.e., elementwise) matrix product,

$$\mathbf{A} * \mathbf{B} = \begin{pmatrix} \mathbf{A}_{11}\mathbf{B}_{11} & \cdots & \mathbf{A}_{1n}\mathbf{B}_{1n} \\ \vdots & \ddots & \vdots \\ \mathbf{A}_{m1}\mathbf{B}_{m1} & \cdots & \mathbf{A}_{mn}\mathbf{B}_{mn} \end{pmatrix}. \tag{4.3}$$

The symbol $\otimes$ denotes the **Kronecker** product, i.e.

$$\mathbf{A} \otimes \mathbf{B} = \begin{pmatrix} \mathbf{A}_{11}\mathbf{B} & \cdots & \mathbf{A}_{1n}\mathbf{B} \\ \vdots & \ddots & \vdots \\ \mathbf{A}_{m1}\mathbf{B} & \cdots & \mathbf{A}_{mn}\mathbf{B} \end{pmatrix}. \tag{4.4}$$

And the symbol $\odot$ denotes the **Khatri-Rao** product (columnwise Kronecker)[37],

$$\mathbf{A} \odot \mathbf{B} = (\mathbf{A}_1 \otimes \mathbf{B}_1 \quad \cdots \quad \mathbf{A}_n \otimes \mathbf{B}_n). \tag{4.5}$$

where $\mathbf{A}_i, \mathbf{B}_i$ are the columns of $\mathbf{A}, \mathbf{B}$ respectively.

The concept of matricizing or unfolding is simply a rearrangement of the entries of $\boldsymbol{T}$ into a matrix. For a three-dimensional array $\boldsymbol{T}$ of size $m \times n \times p$, the notation $\boldsymbol{T}^{(m \times np)}$ represents a matrix of size $m \times np$ in which the n-index runs the fastest over columns and p the slowest. The norm of a tensor, $||\boldsymbol{T}||$, is the same as the Frobenius norm of the matricized array, i.e., the square root of the sum of squares of all its elements.

Nonnegative Rank-k Tensor Decomposition Problem:

$$\min_{x^{(i)}, y^{(i)}, z^{(i)}} ||\boldsymbol{T} - \sum_{i=1}^{r} \boldsymbol{x}^{(i)} \circ \boldsymbol{y}^{(i)} \circ \boldsymbol{z}^{(i)}||, \tag{4.6}$$

subject to:

$$\boldsymbol{x}^{(i)} \geq \mathbf{0}, \boldsymbol{y}^{(i)} \geq \mathbf{0}, \boldsymbol{z}^{(i)} \geq \mathbf{0}$$

where $\boldsymbol{T} \in \mathbb{R}^{m \times n \times p}, \boldsymbol{x}^{(i)} \in \mathbb{R}^m, \boldsymbol{y}^{(i)} \in \mathbb{R}^n, \boldsymbol{z}^{(i)} \in \mathbb{R}^p$.

Note that Equation (4.6) defines matrices $\mathbf{X}$ which is $m \times k$, $\mathbf{Y}$ which is $n \times k$, and $\mathbf{X}$ which is $p \times k$. Also, see Figure 2 which provides an illustration of 3D tensor approximation by a sum of rank one tensors. When the sum is truncated after, say, k terms, it then provides a rank k approximation to the tensor $\boldsymbol{T}$.

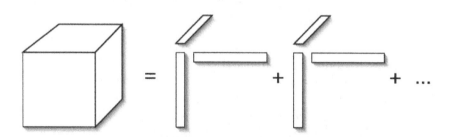

Fig. 2 An illustration of 3-D tensor factorization.

Alternating least squares for NTF

A common approach to solving Equation (4.6) is an alternating least squares (ALS) algorithm [22, 30, 77], due to its simplicity and ability to handle constraints. At each inner iteration, we compute an entire factor matrix while holding all the others fixed.

Starting with random initializations for X, Y and Z, we update these quantities in an alternating fashion using the method of normal equations. The minimization problem involving X in Equation (4.6) can be rewritten in matrix form as a least squares problem:

$$\min_{X} ||T^{(m \times np)} - XC||^2. \tag{4.7}$$

where $T^{(m \times np)} = X(Z \odot Y)^T, C = (Z \odot Y)^T$.

The least squares solution for Equation (4.6) involves the pseudo-inverse of C, which may be computed in a special way that avoids computing $C^T C$ with an explicit C, so the solution to Equation (4.6) is given by

$$X = T^{(m \times np)}(Z \odot Y)(Y^T Y * Z^T Z)^{-1}. \tag{4.8}$$

Furthermore, the product $T^{(m \times np)}(Z \odot Y)$ may be computed efficiently if T is sparse by not forming the Khatri-Rao product $(Z \odot Y)$. Thus, computing X essentially reduces to several matrix inner products, tensor-matrix multiplication of Y and Z into T, and inverting an $R \times R$ matrix.

Analogous least squares steps may be used to update Y and Z. Following is a summary of the complete NTF algorithm.

ALS algorithm for NTF:

---

1. Group x_i's, y_i's and z_i's as columns in $X \in \mathbb{R}_+^{m \times r}, Y \in \mathbb{R}_+^{n \times r}$ and $Z \in \mathbb{R}_+^{p \times r}$ respectively.
2. Initialize X, Y.
   (a) Nonnegative Matrix Factorization of the mean slice,

   $$\min ||A - XY||_F^2. \tag{4.9}$$

   where A is the mean of T across the 3rd dimension.
3. Iterative Tri-Alternating Minimization
   (a) Fix T, X, Y and fit Z by solving a NMF problem in an alternating fashion.

   $$X_{i\rho} \leftarrow X_{i\rho} \frac{(T^{(m \times np)}C)_{i\rho}}{(XC^T C)_{i\rho} + \epsilon}, \quad C = (Z \odot Y) \tag{4.10}$$

   (b) Fix T, X, Z, fit for Y,

   $$Y_{j\rho} \leftarrow Y_{j\rho} \frac{(T^{(m \times np)}C)_{j\rho}}{(YC^T C)_{j\rho} + \epsilon}, \quad C = (Z \odot X) \tag{4.11}$$

   (c) Fix T, Y, Z, fit for X.

   $$Z_{k\rho} \leftarrow Z_{k\rho} \frac{(T^{(m \times np)}C)_{k\rho}}{(ZC^T C)_{k\rho} + \epsilon}, \quad C = (Y \odot X) \tag{4.12}$$

Here ϵ is a small number like 10^{-9} that adds stability to the calculation and guards against introducing a negative number from numerical underflow.

If $\mathcal{T}$ is sparse a simpler computation in the procedure above can be obtained. Each matricized version of $\mathcal{T}$ is a sparse matrix. The matrix C from each step should not be formed explicitly because it would be a large, dense matrix. Instead, the product of a matricized $\mathcal{T}$ with C should be computed specially, exploiting the inherent Kronecker product structure in C so that only the required elements in C need to be computed and multiplied with the nonzero elements of $\mathcal{T}$.

5 Some applications of nonnegativity constraints

5.1 Support vector machines

Support Vector machines were introduced by Vapnik and co-workers [11, 18] theoretically motivated by Vapnik-Chervonenkis theory (also known as VC theory [81, 82]). Support vector machines (SVMs) are a set of related supervised learning methodslearning used for classification and regression. They belong to a family of generalized linear classifiers. They are based on the following idea: input points are mapped to a high dimensional feature space, where a separating hyperplane can be found. The algorithm is chosen in such a way to maximize the distance from the closest patterns, a quantity that is called the margin. This is achieved by reducing the problem to a quadratic programming problem,

$$F(v) = \frac{1}{2}\mathbf{v}^T \mathbf{A} \mathbf{v} + \mathbf{b}^T \mathbf{v}, \quad \mathbf{v} \geq 0. \tag{5.1}$$

Here we assume that the matrix $\mathbf{A}$ is symmetric and semipositive definite. The problem (5.1) is then usually solved with optimization routines from numerical libraries. SVMs have a proven impressive performance on a number of real world problems such as optical character recognition and face detection.

We briefly review the problem of computing the maximum margin hyperplane in SVMs [81]. Let $\{(x_i, y_i)\}_i^N = 1\}$ denote labeled examples with binary class labels $y_i = \pm 1$, and let $K(x_i, x_j)$ denote the kernel dot product between inputs. For brevity, we consider only the simple case where in the high dimensional feature space, the classes are linearly separable and the hyperplane is required to pass through the origin. In this case, the maximum margin hyperplane is obtained by minimizing the loss function:

$$L(\alpha) = -\sum_i \alpha_i + \frac{1}{2}\sum_{ij} \alpha_i \alpha_j y_i y_j K(x_i, x_j), \tag{5.2}$$

subject to the nonnegativity constraints $\alpha_i \geq 0$. Let α^* denote the minimum of equation (5.2). The maximal margin hyperplane has normal vector $w = \sum_i \alpha_i^* y_i x_i$ and satisfies the margin constraints $y_i K(w, x_i) \geq 1$ for all examples in the training set.

The loss function in equation (5.2) is a special case of the non-negative quadratic programming (5.1) with $A_{ij} = y_i y_j K(x_i, x_j)$ and $\mathbf{b}_i = -1$. Thus, the multiplicative updates in the paper [72] are easily adapted to SVMs. This algorithm for training SVMs is known as Multiplicative Margin Maximization (M^3). The algorithm can be generalized to data that is not linearly separable and to separating hyper-planes that do not pass through the origin.

Many iterative algorithms have been developed for nonnegative quadratic programming in general and for SVMs as a special case. Benchmarking experiments have shown that M^3 is a feasible algorithm for small to moderately sized data sets. On the other hand, it does not converge as fast as leading subset methods for large data sets. Nevertheless, the extreme simplicity and convergence guarantees of M^3 make it a useful starting point for experimenting with SVMs.

5.2   Image processing and computer vision

Digital images are represented nonnegative matrix arrays, since pixel intensity values are nonnegative. It is sometimes desirable to process data sets of images represented by column vectors as composite objects in many articulations and poses, and sometimes as separated parts for in, for example, biometric identification applications such as face or iris recognition. It is suggested that the factorization in the linear model would enable the identification and classification of intrinsic "parts" that make up the object being imaged by multiple observations [14, 36, 46, 48]. More specifically, each column x_j of a nonnegative matrix $\mathbf{X}$ now represents m pixel values of one image. The columns w_i of $\mathbf{W}$ are basis elements in R^m. The columns of $\mathbf{H}$, belonging to R^k, can be thought of as coefficient sequences representing the n images in the basis elements. In other words, the relationship

$$\mathbf{x}_j = \sum_{i=1}^{k} w_i h_{ij}, \tag{5.3}$$

can be thought of as that there are standard parts $\mathbf{w}_i$ in a variety of positions and that each image represented as a vector $\mathbf{x}_j$ making up the factor $\mathbf{W}$ of basis elements is made by superposing these parts together in specific ways by a mixing matrix represented by $\mathbf{H}$. Those parts, being images themselves, are necessarily nonnegative. The superposition coefficients, each part being present or absent, are also necessarily nonnegative. A related application to the identification of object materials from spectral reflectance data at different optical wavelengths has been investigated in [61–63].

As one of the most successful applications of image analysis and understanding, face recognition has recently received significant attention, especially during the past few years. Recently, many papers, like [8, 36, 46, 49, 57] have proved that Nonnegative Matrix Factorization (NMF) is a good method to obtain a representation of data using non-negativity constraints. These constraints lead to a part-based representation because they allow only additive, not subtractive, combinations of the original data. Given an initial database expressed by a $n \times m$ matrix $\mathbf{X}$, where each column is an n-dimensional nonnegative vector of the original database (m vectors), it is possible to find two new matrices $\mathbf{W}$ and $\mathbf{H}$ in order to approximate the original matrix

$$X_{i\mu} \approx (WH)_{i\mu} = \sum_{a=1}^{k} W_{ia} H_{a\mu}. \tag{5.4}$$

The dimensions of the factorized matrices $\mathbf{W}$ and $\mathbf{H}$ are $n \times k$ and $k \times m$, respectively. Usually, k is chosen so that $(n+m)k < nm$. Each column of matrix $\mathbf{W}$ contains a basis vector while each column of $\mathbf{H}$ contains the weights needed to approximate the corresponding column in $\mathbf{X}$ using the bases from $\mathbf{W}$.

Other image processing work that uses non-negativity constraint includes the work image restorations. Image restoration is the process of approximating an original image from an observed blurred and noisy image. In image restoration, image formation is modeled as a first kind integral equation which, after discretization, results in a large scale linear system of the form

$$\mathbf{A}\mathbf{x} + \eta = \mathbf{b}. \tag{5.5}$$

The vector $\mathbf{x}$ represents the true image, $\mathbf{b}$ is the blurred noisy copy of $\mathbf{x}$, and η models additive noise, matrix $\mathbf{A}$ is a large ill-conditioned matrix representing the blurring phenomena.

In the absence of information of noise, we can model the image restoration problem as NNLS problem,

$$\min_{\mathbf{x}} \frac{1}{2} \|\mathbf{A}\mathbf{x} - \mathbf{b}\|^2, \\ subject \ \ to \ \ \mathbf{x} \geq 0. \tag{5.6}$$

Thus, we can use NNLS to solve this problem. Experiments show that enforcing a nonnegativity constraint can produce a much more accurate approximate solution, see e.g., [29, 38, 54, 70].

5.3   Text mining

Assume that the textual documents are collected in an *matrix* $\mathbf{Y} = [y_{ij}] \in R^{m \times n}$. Each document is represented by one column in $\mathbf{Y}$. The entry y_{ij} represents the *weight* of one particular *term* i in document j whereas each term could be defined by just one single word or a string of phrases. To enhance discrimination between various documents and to improve retrieval effectiveness, a term-weighting scheme of the form,

$$y_{ij} = t_{ij} g_i d_j, \tag{5.7}$$

is usually used to define $\mathbf{Y}$ [9], where t_{ij} captures the relative importance of term i in document j, g_i weights the overall importance of term i in the entire set of documents, and $d_j = (\sum_{i=1}^{m} t_{ij} g_i)^{-1/2}$ is the scaling factor for normalization. The normalization by d_j per document is necessary, otherwise one could artificially inflate the prominence of document j by padding it with repeated pages or volumes. After the normalization, the columns of $\mathbf{Y}$ are of unit length and usually nonnegative.

The indexing matrix contains lot of information for retrieval. In the context of latent semantic indexing (LSI) application [9, 31], for example, suppose a query represented by a row vector $\mathbf{q}^T = [q_1, ..., q_m] \in R^m$, where q_i denotes the weight of term i in the query $\mathbf{q}$, is submitted. One way to measure how the query $\mathbf{q}$ matches the documents is to calculate the row vector $\mathbf{s}^T = \mathbf{q}^T \mathbf{Y}$ and rank the relevance of documents to $\mathbf{q}$ according to the *scores* in $\mathbf{s}$.

The computation in the LSI application seems to be merely the vector-matrix multiplication. This is so only if $\mathbf{Y}$ is a "reasonable" representation of the relationship between documents and terms. In practice, however, the matrix $\mathbf{Y}$ is never exact. A major challenge in the field has been to represent the indexing matrix and the queries in a more compact form so as to facilitate the computation of the scores [20, 59]. The idea of representing $\mathbf{Y}$ by its nonnegative matrix factorization approximation seems plausible. In this context, the standard parts w_i indicated in (5.3) may be interpreted as subcollections of some "general concepts" contained in these documents. Like images, each document can be thought of as a linear composition of these general concepts. The column-normalized matrix $\mathbf{A}$ itself is a term-concept indexing matrix.

5.4   Environmetrics and chemometrics

In the air pollution research community, one observational technique makes use of the ambient data and source profile data to apportion sources or source categories [34, 39, 68]. The fundamental principle in this model is that mass conservation can be assumed and a mass balance analysis can be used to identify and apportion sources of airborne particulate matter in the atmosphere. For example, it might be desirable to determine a large number of chemical constituents such as elemental concentrations in a number of samples. The relationships between p sources which contribute m chemical species to n samples leads to a mass balance equation

$$y_{ij} = \sum_{k=1}^{p} a_{ik} f_{kj}, \qquad (5.8)$$

where y_{ij} is the elemental concentration of the ith chemical measured in the jth sample, a_{ik} is the gravimetric concentration of the ith chemical in the kth source, and f_{kj} is the airborne mass concentration that the kth source has contributed to the jth sample. In a typical scenario, only values of y_{ij} are observable whereas neither the sources are known nor the compositions of the local particulate emissions are measured. Thus, a critical question is to estimate the number p, the compositions a_{ik}, and the contributions f_{kj} of the sources. Tools that have been employed to analyze the linear model include principal component analysis, factor analysis, cluster analysis, and other multivariate statistical techniques. In this receptor model, however, there is a physical constraint imposed upon the data. That is, the source compositions a_{ik} and the source contributions f_{kj} must all be nonnegative. The identification and apportionment problems thus become a nonnegative matrix factorization problem for the matrix $\mathbf{Y}$.

5.5   Speech recognition

Stochastic language modeling plays a central role in large vocabulary speech recognition, where it is usually implemented using the n-gram paradigm. In a typical application, the purpose of an n-gram language model may be to constrain the acoustic analysis, guide the search through various (partial) text hypotheses, and/or contribute to the determination of the final transcription.

In language modeling one has to model the probability of occurrence of a predicted word given its history $Pr(w_n|\mathbf{H})$. N-gram based Language Models have been used successfully in Large Vocabulary Automatic Speech Recognition Systems. In this model, the word history consists of the $N-1$ immediately preceding words. Particularly, tri-gram language models ($Pr(w_n|w_{n-1}; w_{n-2})$) offer a good compromise between modeling power and complexity. A major weakness of these models is the inability to model word dependencies beyond the span of the n-grams. As such, n-gram models have limited semantic modeling ability. Alternate models have been proposed with the aim of incorporating long term dependencies into the modeling process. Methods such as word trigger models, high-order n-grams, cache models, etc., have been used in combination with standard n-gram models.

One such method, a *Latent Semantic Analysis* based model has been proposed [5]. A word-document occurrence matrix $\mathbf{X}_{m \times n}$ is formed (m = size of the vocabulary, n = number of documents), using a training corpus explicitly segmented into a collection of documents. A Singular Value Decomposition $\mathbf{X} = \mathbf{USV}^T$ is performed to obtain a low dimensional linear space $\mathcal{S}$, which is more convenient to perform tasks such as word and document clustering, using an appropriate metric. Bellegarda [5] gave the detailing explanation about this method.

In the paper [56], Novak and Mammone introduce a new method with NMF. In addition to the non-negativity, another property of this factorization is that the columns of $\mathbf{W}$ tend to represent groups of associated words. This property suggests that the columns of $\mathbf{W}$ can be interpreted as conditional word probability distributions, since they satisfy the conditions of a probability distribution by the definition. Thus the matrix $\mathbf{W}$ describes a hidden document space $\mathcal{D} = \{d_j\}$ by providing conditional distributions $\mathbf{W} = \mathbf{P}(w_i|d_j)$. The task is to find a matrix $\mathbf{W}$, given the word document count matrix $\mathbf{X}$. The second term of the factorization, matrix $\mathbf{H}$, reflects the properties of the explicit segmentation of the training corpus into individual documents. This information is not of interest in the context of Language Modeling. They provide an experimental result where the NMF method results in a perplexity reduction of 16% on a database of biology lecture transcriptions.

5.6   Spectral unmixing by NMF and NTF

Here we discuss applications of NMF and NTF to numerical methods for the classification of remotely sensed objects. We consider the identification of space satellites from non-imaging data such as spectra of visible and NIR range, with different spectral resolutions and in the presence of noise and atmospheric turbulence. (See, e.g., [61] or [62, 63].) This is the research area of space object identification (SOI).

A primary goal of using remote sensing image data is to identify materials present in the object or scene being imaged and quantify their abundance estimation, i.e., to determine concentrations of different signature spectra present in pixels. Also, due to the large quantity of data usually encountered in hyperspectral datasets, compressing the data is becoming increasingly important. In this section we discuss the use of MNF and NTF to reach these major goals: material identification, material abundance estimation, and data compression.

For safety and other considerations in space, non-resolved space object characterization is an important component of Space Situational Awareness. The key problem in non-resolved space object characterization is to use spectral reflectance data to gain knowledge regarding the physical properties (e.g., function, size, type, status change) of space objects that cannot be spatially resolved with telescope technology. Such objects may include geosynchronous satellites, rocket bodies, platforms, space debris, or nano-satellites. rendition of a JSAT type satellite in a 36,000 kilometer high synchronous orbit around the Earth. Even with adaptive optics capabilities, this object is generally not resolvable using ground-based telescope technology.

Fig. 3 Artist rendition of a JSAT satellite. Image obtained from the Boeing Satellite Development Center.

Spectral reflectance data of a space object can be gathered using ground-based spectrometers and contains essential information regarding the make up or types of materials comprising the object. Different materials such as aluminum, mylar, paint, etc. possess characteristic wavelength-dependent absorption features, or spectral *signatures*, that mix together in the spectral reflectance measurement of an object. Figure 4 shows spectral signatures of four materials typically used in satellites, namely, aluminum, mylar, white paint, and solar cell.

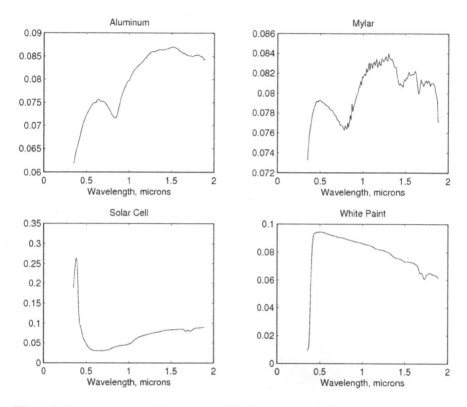

Fig. 4 Laboratory spectral signatures for aluminum, mylar, solar cell, and white paint. For details see [63].

The objective is then, given a set of spectral measurements or traces of an object, to determine i) the type of constituent materials and ii) the proportional amount in which these materials appear. The first problem involves the detection of material spectral signatures or *endmembers* from the spectral data. The second problem involves the computation of corresponding proportional amounts or *fractional abundances*. This is known as the *spectral unmixing* problem in the hyperspectral imaging community.

Recall that in In Nonnegative Matrix Factorization (NMF), an $m \times n$ (nonnegative) mixed data matrix $\mathbf{X}$ is approximately factored into a product of two nonnegative rank-k matrices, with k small compared to m and n, $\mathbf{X} \approx \mathbf{WH}$. This factorization has the advantage that $\mathbf{W}$ and $\mathbf{H}$ can provide a physically realizable representation of the mixed data, see e.g. [61]. Two sets of factors, one as endmembers and the other as fractional abundances, are optimally fitted simultaneously. And due to reduced sizes of factors, data compression, spectral signature identification of constituent materials, and determination of their corresponding fractional abundances, can be fulfilled at the same time.

Spectral reflectance data of a space object can be gathered using ground-based spectrometers, such as the SPICA system located on the 1.6 meter Gemini telescope and the ASIS system located on the 3.67 meter telescope at the Maui Space Surveillance Complex (MSSC), and contains essential information regarding the make up or types of materials comprising the object. Different materials, such as aluminum, mylar, paint, plastics and solar cell, possess characteristic wavelength-dependent absorption features, or spectral *signatures*, that mix together in the spectral reflectance measurement of an object. A new spectral imaging sensor, capable of collecting hyperspectral images of space objects, has been installed on the 3.67 meter Advanced Electrocal-optical System (AEOS) at the MSSC. The AEOS Spectral Imaging Sensor (ASIS) is used to collect adaptive optics compensated spectral images of astronomical objects and satellites. See Figure 4 for a simulated hyperspectral image of the Hubble Space Telescope similar to that collected by ASIS.

Fig. 5 A blurred and noisy simulated hyperspectral image above the original simulated image of the Hubble Space Telescope representative of the data collected by the Maui ASIS system.

In [85] and [86] Zhang, et al. develop NTF methods for identifying space objects using hyperspectral data. Illustrations of material identification, material abundance estimation, and data compression are demonstrated for data similar to that shown in Figure 5.

6   Summary

We have outlined some of what we consider the more important and interesting problems for enforcing nonnegativity constraints in numerical analysis. Special emphasis has been placed nonnegativity constraints in least squares computations in numerical linear algebra and in nonlinear optimization. Techniques involving nonnegative low-rank matrix and tensor factorizations and their many applications were also given. This report also includes an effort toward a literature survey of the various algorithms and applications of nonnegativity constraints in numerical analysis. As always, such an overview is certainly incomplete, and we apologize for omissions. Hopefully, this work will inform the reader about the importance of nonnegativity constraints in many problems in numerical analysis, while pointing toward the many advantages of enforcing nonnegativity in practical applications.

References

1. R.B. Bapat, T.E.S. Raghavan, *Nonnegative Matrices and Applications*, Cambridge University Press, UK, 1997.
2. M.S. Bartlett, J.R. Movellan, and T.J. Sejnowski, Face recognition by independent component analysis, *IEEE Trans. Neural Networks*, **13**, No. 6, (2002) 1450–1464.
3. P. Belhumeur, J. Hespanha, and D. Kriegman, Eigenfaces vs. Fisherfaces: Recognition Using Class Specific Linear Projection, *IEEE PAMI*, **19**, No. 7, (1997) 711–720.
4. S. Bellavia, M. Macconi, and B. Morini, An interior point newton-like method for nonnegative least squares problems with degenerate solution, *Numerical Linear Algebra with Applications*, **13**, (2006) 825–846.
5. J.R. Bellegarda, A multispan language modelling framework for large vocabulary speech recognition, *IEEE Transactions on Speech and Audio Processing*, **6** No. 5, (1998) 456–467.
6. A. Berman and R. Plemmons, Rank factorizations of nonnegative matrices, Problems and Solutions, 73-14 (Problem), SIAM Rev., **15** (1973) 655.
7. A. Berman, R. Plemmons, *Nonnegative Matrices in the Mathematical Sciences*, Academic Press, NY, 1979. Revised version in SIAM Classics in Applied Mathematics, Philadelphia, 1994.
8. M. Berry, M. Browne, A. Langville, P. Pauca, and R. Plemmons, Algorithms and applications for approximate nonnegative matrix factorization, *Computational Statistics and Data Analysis*, **52**, (2007) 155–173. Preprint available at http://www.wfu.edu/~plemmons
9. M.W. Berry, *Computational Information Retrieval*, SIAM, Philadelphia, 2000.
10. D. Bertsekas, *Nonlinear Programming*, Athena Scientific, Belmont, MA., 1999.
11. B. Boser, I. Guyon, and V. Vapnik, *A training algorithm for optimal margin classifiers*, Fifth Annual Workshop on Computational Learning Theory, ACM Press, 1992.
12. R. Bro, S. de Jong, A fast non-negativity-constrained least squares algorithm, *Journal of Chemometrics*, **11**, No. 5, (1997) 393–401.
13. J. Cantarella, M. Piatek, Tsnnls: A solver for large sparse least squares problems with non-negative variables, *ArXiv Computer Science e-prints*, 2004.
14. X. Chen, L. Gu, S.Z. Li, and H.J. Zhang, Learning representative local features for face detection, in *IEEE Conference on Computer Vision and Pattern Recognition*, **1**, (2001) 1126–1131.

15. M.T. Chu, F. Diele, R. Plemmons, S. Ragni, *Optimality, computation, and interpretation of nonnegative matrix factorizations*, preprint. Available at: http://www.wfu.edu/~plemmons

16. M.T. Chu and R.J. Plemmons, Nonnegative matrix factorization and applications, Appeared in *IMAGE, Bulletin of the International Linear Algebra Society*, **34**, (2005) 2–7. Available at: http://www.wfu.edu/~plemmons

17. A. Cichocki, R. Zdunek, and S. Amari, Hierarchical ALS Algorithms for Nonnegative Matrix and 3D Tensor Factorization, In: Independent Component Analysis, ICA07, London, UK, September 9-12, 2007, *Lecture Notes in Computer Science*, Vol. LNCS 4666, Springer, (2007) 169–176.

18. C. Cortes, V. Vapnik, Support Vector networks, *Machine Learning*, **20**, (1995) 273–297.

19. A. Dax, On computational aspects of bounded linear least squares problems, *ACM Trans. Math. Softw.*, **17**, (1991) 64–73.

20. I.S. Dhillon, D.M. Modha, Concept decompositions for large sparse text data using clustering, *Machine Learning J.*, **42**, (2001) 143–175.

21. C. Ding, X. He, and H. Simon, *On the equivalence of nonnegative matrix factorization and spectral clustering*, Proceedings of the Fifth SIAM International Conference on Data Mining, Newport Beach, CA, 2005.

22. N.K.M. Faber, R. Bro, and P.K. Hopke, Recent developments in CANDECOMP/PARAFAC algorithms: a critical review, *Chemometr. Intell. Lab.*, **65**, No. 1, (2003) 119–137.

23. V. Franc, V. Hlaváč, and M. Navara, *Sequential coordinate-wise algorithm for nonnegative least squares problem*, Research report CTU-CMP-2005-06, Center for Machine Perception, Czech Technical University, Prague, Czech Republic, February 2005.

24. P.E. Gill, W. Murray and M.H. Wright, *Practical Optimization*, Academic, London, 1981.

25. A.A. Giordano, F.M. Hsu, *Least Square Estimation With Applications To Digital Signal Processing*, John Wiley & Sons, 1985.

26. L. Grippo, M. Sciandrone, On the convergence of the block nonlinear Gauss-Seidel method under convex constraints, *Oper. Res. Lett.*, **26**, No. 3, (2000) 127–136.

27. D. Guillamet, J. Vitrià, *Classifying faces with non-negative matrix factorization*, Accepted CCIA 2002, Castelló de la Plana, Spain.

28. D. Guillamet, J. Vitrià, Non-negative matrix factorization for face recognition, *Lecture Notes in Computer Science*, **2504**, (2002) 336–344.

29. M. Hanke, J.G. Nagy and C.R. Vogel, Quasi-newton approach to nonnegative image restorations, *Linear Algebra Appl.*, **316**, (2000) 223–236.

30. R.A. Harshman, *Foundations of the PARAFAC procedure: models and conditions for an "explantory" multi-modal factor analysis*, UCLA working papers in phonetics, Vol. 16, pp. 1–84, 1970.

31. T. Hastie, R. Tibshirani, and J. Friedman, *The Elements of Statistical Learning: Data Mining, Inference, and Prediction*, Springer-Verlag, New York, 2001.

32. N.-D. Ho, *Nonnegative Matrix Factorization Algorithms and Applications*, PhD thesis, Univ. Catholique de Louvain, June 2008. (Available from edoc.bib.ucl.ac.be:81/ETD-db/collection/available/BelnUcetd-06052008-235205/).

33. P.K. Hopke, *Receptor Modeling in Environmental Chemistry*, Wiley and Sons, New York, 1985.

34. P.K. Hopke, *Receptor Modeling for Air Quality Management*, Elsevier, Amsterdam, Netherlands, 1991.

35. P.O. Hoyer, *Nonnegative sparse coding, neural networks for signal processing XII*, Proc. IEEE Workshop on Neural Networks for Signal Processing, Martigny, 2002.

36. P. Hoyer, Nonnegative matrix factorization with sparseness constraints, *J. of Mach. Leanring Res.*, **5**, (2004) 1457–1469.

37. C.G. Khatri, C.R. Rao, Solutions to some functional equations and their applications to. characterization of probability distributions, *Sankhya*, **30**, (1968) 167–180.

38. B. Kim, *Numerical optimization methods for image restoration*, Ph.D. thesis, Stanford University, 2002.

39. E. Kim, P.K. Hopke, E.S. Edgerton, Source identification of Atlanta aerosol by positive matrix factorization, *J. Air Waste Manage. Assoc.*, **53**, (2003) 731–739.

40. H. Kim, H. Park, Sparse non-negative matrix factorizations via alternating nonnegativity-constrained least squares for microarray data analysis, *Bioinformatics*, **23**, No. 12, (2007) 1495–1502.

41. D. Kim, S. Sra, and I.S. Dhillon, *A new projected quasi-newton approach for the nonnegative least squares problem*, Dept. of Computer Sciences, The Univ. of Texas at Austin, Technical Report # TR-06-54, Dec. 2006.

42. D. Kim, S. Sra, and I.S. Dhillon, Fast newton-type methods for the least squares nonnegative matrix approximation problem, *Statistical Analysis and Data Mining*, **1**, No. 1, (2008) 38-51.

43. S. Kullback, and R. Leibler, On information and sufficiency, *Annals of Mathematical Statistics*, **22** (1951) 79–86.

44. A. Langville, C. Meyer, R. Albright, J. Cox, D. Duling, *Algorithms, initializations, and convergence for the nonnegative matrix factorization*, NCSU Technical Report Math 81706, 2006.

45. C.L. Lawson and R.J. Hanson, *Solving Least Squares Problems*, Prentice-Hall, 1987.

46. D. Lee and H. Seung, Learning the parts of objects by non-negative matrix factorization, *Nature*, **401**, (1999) 788–791.

47. D. Lee and H. Seung, Algorithms for nonnegative matrix factorization, *Advances in Neural Information Processing Systems*, **13**, (2001) 556–562.

48. S.Z. Li, X.W. Hou and H.J. Zhang, *Learning spatially localized, parts-based representation*, *IEEE* Conference on Computer Vision and Pattern Recognition, (2001) 1–6.

49. C.J. Lin, Projected gradient methods for non-negative matrix factorization, *Neural Computation*, **19**, No. 10, (2007) 2756–2779.

50. W. Liu, J. Yi, *Existing and new algorithms for nonnegative matrix factorization*, University of Texas at Austin, 2003, report.

51. P. Matstoms, *snnls: a matlab toolbox for Solve sparse linear least squarest problem with nonnegativity constraints by an active set method*, 2004, available at http://www.math.liu.se/~milun/sls/.

52. A. Mazer, M. Martin, M. Lee and J. Solomon, Image processing software for imaging spectrometry data analysis, *Remote Sensing of Environment*, **24**, (1988) 201–220.

53. J.J. Moré, G. Toraldo, On the solution of large quadratic programming problems with bound constraints, *SIAM Journal on Optimization*, **1**, No. 1, (1991) 93–113.

54. J.G. Nagy, Z. Strakoš, *Enforcing nonnegativity in image reconstruction algorithms*, in Mathematical Modeling, Estimation, and Imaging, 4121, David C. Wilson, et al, (eds.), (2000) 182–190.

55. J. Nocedal, S. Wright, *Numerical Optimization*, Springer, Berlin, 2006.

56. M. Novak, R. Mammone, *Use of non-negative matrix factorization for language model adaptation in a lecture transcription task*, *IEEE* Workshop on ASRU 2001, pp. 190–193, 2001.

57. P. Paatero and U. Tapper, Positive matrix factorization – a nonnegative factor model with optimal utilization of error-estimates of data value, *Environmetrics*, **5**, (1994) 111–126.

58. P. Paatero, The multilinear engine – a table driven least squares program for solving mutilinear problems, including the n-way parallel factor analysis model, *J. Comput. Graphical Statist.*, **8**, No. 4, (1999) 854–888.

59. H. Park, M. Jeon, J.B. Rosen, *Lower dimensional representation of text data in vector space based information retrieval*, in Computational Information Retrieval, ed. M. Berry, Proc. Comput. Inform. Retrieval Conf., SIAM, (2001) 3–23.

60. V.P. Pauca, F. Shahnaz, M.W. Berry, R.J. Plemmons, *Text mining using nonnegative matrix factorizations*, In Proc. SIAM Inter. Conf. on Data Mining, Orlando, FL, April 2004.

61. P. Pauca, J. Piper, and R. Plemmons, Nonnegative matrix factorization for spectral data analysis, *Lin. Alg. Applic.*, **416**, No. 1, (2006) 29–47.

62. P. Pauca, J. Piper, R. Plemmons, M. Giffin, *Object characterization from spectral data using nonnegative factorization and information theory*, Proc. AMOS Technical Conf., Maui HI, September 2004. Available at http://www.wfu.edu/~plemmons

63. P. Pauca, R. Plemmons, M. Giffin and K. Hamada, *Unmixing spectral data using nonnegative matrix factorization*, Proc. AMOS Technical Conference, Maui, HI, September 2004. Available at http://www.wfu.edu/~plemmons

64. E. Polak, *Computational Methos in Optimization: A Unified Approach*, Academic Press, New York, 1971.

65. L.F. Portugal, J.J. Judice, and L.N. Vicente, A comparison of block pivoting and interior-point algorithms for linear least squares problems with nonnegative variables, *Mathematics of Computation*, **63**, No. 208, (2004) 625–643.

66. M. Powell, An efficient method for finding the minimum of a function of several variables without calculating derivatives, *Comput. J.* **7**, (1964) 155-162.

67. M. Powell, On Search Directions For Minimization, *Math. Programming* **4**, (1973) 193–201.

68. Z. Ramadan, B. Eickhout, X. Song, L.M.C. Buydens, P.K. Hopke, Comparison of positive matrix factorization and multilinear engine for the source apportionment of particulate pollutants, *Chemometrics and Intelligent Laboratory Systems*. **66**, (2003) 15–28.

69. R. Ramath, W. Snyder, and H. Qi, *Eigenviews for object recognition in multi-spectral imaging systems*, 32nd Applied Imagery Pattern Recognition Workshop, Washington D.C., pp. 33–38, 2003.

70. M. Rojas, T. Steihaug, *Large-Scale optimization techniques for nonnegative image restorations*, Proceedings of SPIE, 4791: 233-242, 2002.

71. K. Schittkowski, The numerical solution of constrained linear least-squares problems, *IMA Journal of Numerical Analysis*, **3**, (1983) 11–36.

72. F. Sha, L.K. Saul, D.D. Lee, *Multiplicative updates for large margin classifiers*, Technical Report MS-CIS-03-12, Department of Computer and Information Science, University of Pennsylvania, 2003.

73. A. Shashua and T. Hazan, *Non-negative tensor factorization with applications to statistics and computer vision*, Proceedings of the 22nd International Conference on Machine Learning, Bonn, Germany, pp. 792–799, 2005.

74. A. Shashua and A. Levin, *Linear image coding for regression and classification using the tensor-rank principal*, Proceedings of the *IEEE* Conference on Computer Vision and Pattern Recognition, 2001.

75. L. Shure, *Brief History of Nonnegative Least Squares in MATLAB*, Blog available at: http://blogs.mathworks.com/loren/2006/ .

76. N. Smith, M. Gales, Speech recognition using SVMs, in *Advances in Neural and Information Processing Systems*, **14**, Cambridge, MA, 2002, MIT Press.

77. G. Tomasi, R. Bro, PARAFAC and missing values, *Chemometr. Intell. Lab.*, **75**, No. 2, (2005) 163–180.

78. M.A. Turk, A.P. Pentland, Eigenfaces for recognition, *Cognitive Neuroscience*, **3**, No. 1, (1991) 71-86.

79. M.A. Turk, A.P. Pentland, *Face recognition using eigenfaces*, Proc. *IEEE* Conference on Computer Vision and Pattern Recognition, Maui, Hawaii, 1991.

80. M.H. van Benthem, M.R. Keenan, Fast algorithm for the solution of large-scale non-negativity constrained least squares problems, *Journal of Chemometrics*, **18**, (2004) 441–450.

81. V. Vapnik, *Statistical Learning Theory*, Wiley, 1998.

82. V. Vapnik, *The Nature of Statistical Learning Theory*, Springer-Verlag, 1999.

83. R.S. Varga, *Matrix Iterative Analysis*, Prentice-Hall, Englewood Cliffs, NJ, 1962.

84. W. Zangwill, Minimizing a function without calculating derivatives. *Comput. J.*, **10**, (1967) 293–296.

85. P. Zhang, H. Wang, R. Plemmons, and P. Pauca, *Spectral unmixing using nonnegative tensor factorization*, Proc. ACM, Conference, Winston-Salem, NC, March 2007.

86. P. Zhang, H. Wang, R. Plemmons, and P. Pauca, Hyperspectral Data Analysis: A Space Object Material Identification Study, *Journal of the Optical Soc. Amer.*, Series A, **25**, (2008) 3001–3012.

On nonlinear optimization since 1959

M. J. D. Powell

Department of Applied Mathematics and Theoretical Physics,
Centre for Mathematical Sciences,
Wilberforce Road, Cambridge CB3 0WA, England.
mjdp@damtp.cam.ac.uk

Abstract. This view of the development of algorithms for nonlinear optimization is based on the research that has been of particular interest to the author since 1959, including several of his own contributions. After a brief survey of classical methods, which may require good starting points in order to converge successfully, the huge impact of variable metric and conjugate gradient methods is addressed. It encouraged the use of penalty and barrier functions for expressing constrained calculations in unconstrained forms, which are introduced briefly, including the augmented Lagrangian method. Direct methods that make linear approximations to constraints became more popular in the late 1970s, especially sequential quadratic programming, which receives attention too. Sometimes the linear approximations are satisfied only if the changes to the variables are so large that the approximations become unsuitable, which stimulated the development of trust region techniques that make partial corrections to the constraints. That work is also introduced, noting that quadratic models of the objective or Lagrange function do not have to be convex. We consider the sequence of models that is given by the symmetric Broyden updating formula in unconstrained optimization, including the case when first derivatives are not available. The emphasis of the paper is on algorithms that can be applied without good initial estimates of the variables.

1 Earlier algorithms

The year 1959 is stated in the title of this paper, because Davidon (1959) published then the report that describes his variable metric method for the unconstrained minimization of a general differentiable function, $F(\underline{x})$, $\underline{x} \in \mathbb{R}^n$, all underlined symbols being vectors. That work provides many of the ideas and techniques that are fundamental to later developments, especially the construction and accumulation of useful second derivative information from changes in first derivatives that become available as the iterations of the calculation proceed. The second derivative information is held in a positive definite matrix, which gives a downhill search direction whenever the gradient $\underline{\nabla}F(\underline{x})$ is nonzero. Thus an initial vector of variables that is close to the solution is not required, and usually the rate of convergence of the iterations is superlinear. It is a fortunate coincidence that I started my research on numerical analysis in 1959. There-fore, beginning in Section 2, a personal view of major advances in nonlinear optimization during my career is presented.

First we recall some classical foundations of optimization, beginning with Newton's method for solving the nonlinear system of equations $\underline{f}(\underline{x})=0$, where $\underline{f}$ is a continuously differentiable function from $\mathbb{R}^n$ to $\mathbb{R}^n$. For any $\underline{x}_k \in \mathbb{R}^n$, let $J(\underline{x}_k)$ be the Jacobian matrix that has the elements

$$[J(\underline{x}_k)]_{ij} = df_i(\underline{x}_k) / dx_j, \qquad 1 \le i, j \le n. \tag{1.1}$$

Then the first order Taylor series provides the approximation

$$\underline{f}(\underline{x}_k + \underline{d}_k) \approx \underline{f}(\underline{x}_k) + J(\underline{x}_k)\,\underline{d}_k, \qquad \underline{d}_k \in \mathbb{R}^n. \tag{1.2}$$

Newton's method is based on the remark that, if $\underline{d}_k$ is defined by equating the right hand side of this expression to zero, then $\underline{x}_k + \underline{d}_k$ may be a good estimate of a vector that satisfies $\underline{f}(\underline{x}_k + \underline{d}_k)=0$. Indeed, given a starting vector $\underline{x}_1 \in \mathbb{R}^n$, the formula

$$\underline{x}_{k+1} = \underline{x}_k - J(\underline{x}_k)^{-1}\,\underline{f}(\underline{x}_k), \qquad k=1,2,3,\ldots, \tag{1.3}$$

is applied, assuming every $J(\underline{x}_k)$ is nonsingular. It is well known that, if $\underline{x}^*$ satisfies $\underline{f}(\underline{x}^*)=0$ and if $J(\underline{x}^*)$ is nonsingular, then $\underline{x}_k$ converges at a superlinear rate to $\underline{x}^*$ as $k \to \infty$, provided that $\underline{x}_1$ is sufficiently close to $\underline{x}^*$.

It happens often in practice, however, that such a starting point $\underline{x}_1$ is not available. Then it is highly useful to employ $\underline{d}_k = -J(\underline{x}_k)^{-1}\underline{f}(\underline{x}_k)$ as a search direction, letting $\underline{x}_{k+1}$ be the vector

$$\underline{x}_{k+1} = \underline{x}_k + \alpha_k\,\underline{d}_k \tag{1.4}$$

for some choice of $\alpha_k > 0$. A usual way of helping convergence is to seek a value of α_k that provides the reduction $\|\underline{f}(\underline{x}_{k+1})\| < \|\underline{f}(\underline{x}_k)\|$ in the Euclidean norm of $\underline{f}$. This strict reduction can be achieved whenever $J(\underline{x}_k)$ is nonsingular and $\|\underline{f}(\underline{x}_k)\|$ is nonzero. One way of establishing this property begins with the remark that the first derivatives at $\alpha=0$ of the functions $\|\underline{f}(\underline{x}_k + \alpha\,\underline{d}_k)\|^2$, $\alpha \in \mathbb{R}$, and $\phi(\alpha) = \|\underline{f}(\underline{x}_k) + \alpha\,J(\underline{x}_k)\,\underline{d}_k\|^2$, $\alpha \in \mathbb{R}$, are the same due to the use of the first order Taylor series. Moreover, $\phi(\alpha)$, $\alpha \in \mathbb{R}$, is a nonnegative quadratic that takes the values $\phi(0) = \|\underline{f}(\underline{x}_k)\|^2$ and $\phi(1)=0$. Thus we deduce the required condition

$$\left[\frac{d}{d\alpha} \|\underline{f}(\underline{x}_k + \alpha\,\underline{d}_k)\|^2\right]_{\alpha=0} = \phi'(0) = -2\,\|\underline{f}(\underline{x}_k)\|^2 < 0. \tag{1.5}$$

Probably this enhancement of Newton's method is also classical. It is easy to show that the line searches may fail to provide $\|\underline{f}(\underline{x}_k)\| \to 0$ as $k \to \infty$, by picking a system of equations that does not have a solution.

Let every α_k in the method of the last paragraph be the value of α that minimizes $\|\underline{f}(\underline{x}_k + \alpha\,\underline{d}_k)\|^2$, $\alpha \ge 0$. It is possible for each iteration to be well-defined, and for $\underline{x}_k$, $k=1,2,3,\ldots$, to converge to a limit $\underline{x}^*$ where the gradient of the function $F(\underline{x}) = \|\underline{f}(\underline{x})\|^2$, $\underline{x} \in \mathbb{R}^n$, is nonzero, but of course $J(\underline{x}^*)$ is singular (Powell, 1970). Then the search direction $\underline{d}_k$ tends to be orthogonal to $\underline{\nabla}F(\underline{x}_k)$ as $k \to \infty$, which is unwelcome when seeking the least value of a differentiable function F.

Such orthogonality is avoided as much as possible in the steepest descent method for minimizing $F(\underline{x})$, $\underline{x} \in \mathbb{R}^n$, where F is now any continuously differentiable function from $\mathbb{R}^n$ to $\mathbb{R}$. The k-th iteration sets $\underline{d}_k = -\underline{\nabla}F(\underline{x}_k)$, where $\underline{x}_1$ is given and where $\underline{x}_k$, $k \geq 2$, is provided by the previous iteration. Termination occurs if $\|\underline{d}_k\|$ is zero or acceptably small, but otherwise a positive step-length α_k is sought, in order to apply formula (1.4). Typically, values of α_k that are too long or too short are avoided by imposing the conditions

$$\left.\begin{array}{r}F(\underline{x}_k + \alpha_k \underline{d}_k) \leq F(\underline{x}_k) + c_1 \alpha_k \underline{d}_k^T \underline{\nabla}F(\underline{x}_k) \\ \underline{d}_k^T \underline{\nabla}F(\underline{x}_k + \alpha_k \underline{d}_k) \geq c_2 \underline{d}_k^T \underline{\nabla}F(\underline{x}_k)\end{array}\right\}, \qquad (1.6)$$

where c_1 and c_2 are prescribed constants that satisfy $0 < c_1 < 0.5$ and $c_1 < c_2 < 1$. Termination occurs too if, in the search for α_k, it is found that F is not bounded below. This method has a very attractive convergence property, namely that, if the number of iterations is infinite and if the points $\underline{x}_k$ remain in a bounded region of $\mathbb{R}^n$, then the sequence of gradients $\underline{\nabla}F(\underline{x}_k)$ tends to zero as $k \to \infty$.

Often in practice, however, the steepest descent method is intolerably slow. For example, we let m and M be positive constants and we apply the method to the quadratic function

$$F(\underline{x}) = m x_1^2 + M x_2^2, \qquad \underline{x} \in \mathbb{R}^2, \qquad (1.7)$$

starting at the point $\underline{x}_1 = (M, m)^T$. Further, we satisfy the line search conditions (1.6) by letting α_k provide the least value of $F(\underline{x}_{k+1}) = F(\underline{x}_k + \alpha_k \underline{d}_k)$ on every iteration. A simple calculation shows that $\underline{x}_{k+1}$ has the components $M\theta^k$ and $m(-\theta)^k$, $k = 1, 2, 3, \ldots$, where $\theta = (M-m)/(M+m)$. Thus, if $\nabla^2 F(\underline{x}^*)$ is very ill-conditioned, then a large number of iterations may be required to obtain a vector of variables that is close enough to the solution $\underline{x}^* = 0$. This slow convergence occurs for most starting points $\underline{x}_1$, but our choice of $\underline{x}_1$ simplifies the analytic derivation of $\underline{x}_{k+1}$.

The classical way of achieving a superlinear rate of convergence when minimizing a twice continuously differentiable function $F(\underline{x})$, $\underline{x} \in \mathbb{R}^n$, is to apply the Newton–Raphson algorithm. In its basic form it is identical to Newton's method for calculating $\underline{x} \in \mathbb{R}^n$ that solves the nonlinear system $\underline{\nabla}F(\underline{x}) = 0$. Putting $\underline{f} = \underline{\nabla}F$ in the definition (1.1) gives a symmetric Jacobian matrix with the elements

$$[G(\underline{x}_k)]_{ij} = d^2 F(\underline{x}_k) / dx_i dx_j, \qquad 1 \leq i, j \leq n, \qquad (1.8)$$

so equation (1.3) takes the form

$$\underline{x}_{k+1} = \underline{x}_k - G(\underline{x}_k)^{-1} \underline{\nabla}F(\underline{x}_k), \qquad k = 1, 2, 3, \ldots. \qquad (1.9)$$

The line search version (1.4), with $\underline{d}_k = -G(\underline{x}_k)^{-1}\underline{\nabla}F(\underline{x}_k)$, can help convergence sometimes when $\underline{x}_1$ is not sufficiently close to the optimal vector of variables $\underline{x}^*$. Then, as in the given extension to Newton's method, one may seek a step-length α_k that provides the reduction $\|\underline{\nabla}F(\underline{x}_{k+1})\| < \|\underline{\nabla}F(\underline{x}_k)\|$. This approach is objectionable, however, because trying to solve $\underline{\nabla}F(\underline{x}) = 0$ can be regarded as seeking a stationary point of F without paying any attention to minimization. Therefore it may be more suitable to let α_k be an estimate of the value of α that minimizes $F(\underline{x}_k + \alpha\,\underline{d}_k)$, $\alpha \in \mathbb{R}$, but this minimum may occur at $\alpha = 0$, even if $\underline{\nabla}F(\underline{x}_k)$ is nonzero.

The remarks of this section have exposed some major disadvantages of classical methods for optimization. Thus we may be able to appreciate better the gains that have been achieved since 1959.

2   Two major advances in unconstrained optimization

I was fortunate in 1962 to obtain a copy of the report of Davidon (1959), after finding a reference to it in a monograph. The report describes an algorithm for unconstrained minimization, which I programmed for a Ferranti Mercury computer, in order to try some numerical experiments. The results were staggering, especially the minimization of a periodic function $F(\underline{x})$, $\underline{x} \in \mathbb{R}^n$, with 100 variables, although problems with $n = 20$ were considered large at that time. The k-th iteration requires a vector of variables $\underline{x}_k$, an $n \times n$ positive definite symmetric matrix H_k, and the gradient $\underline{\nabla}F(\underline{x}_k)$, which is available from the $(k-1)$-th iteration for $k \geq 2$. The sequence of iterations is terminated if $\|\underline{\nabla}F(\underline{x}_k)\|$ is sufficiently small, but otherwise formula (1.4) gives the next vector of variables, $\underline{d}_k$ being the search direction $\underline{d}_k = -H_k\underline{\nabla}F(\underline{x}_k)$, which has the downhill property $\underline{d}_k^T\underline{\nabla}F(\underline{x}_k) < 0$, and α_k being a step-length that satisfies the conditions (1.6), usually with $|\underline{d}_k^T\underline{\nabla}F(\underline{x}_k + \alpha_k\,\underline{d}_k)|$ much less than $|\underline{d}_k^T\underline{\nabla}F(\underline{x}_k)|$. Finally, the iteration replaces H_k by the matrix

$$H_{k+1} = H_k - \frac{H_k\underline{\gamma}_k\underline{\gamma}_k^T H_k}{\underline{\gamma}_k^T H_k \underline{\gamma}_k} + \frac{\underline{\delta}_k\underline{\delta}_k^T}{\underline{\delta}_k^T\underline{\gamma}_k}, \tag{2.1}$$

where $\underline{\delta}_k = \underline{x}_{k+1} - \underline{x}_k$, where $\underline{\gamma}_k = \underline{\nabla}F(\underline{x}_{k+1}) - \underline{\nabla}F(\underline{x}_k)$, and where the superscript "T" distinguishes a row vector from a column vector. The positive definiteness of H_{k+1} is inherited from H_k, because the second of the conditions (1.6) implies $\underline{\delta}_k^T\underline{\gamma}_k > 0$.

Davidon (1959) explains that, if the objective function F is strictly convex and quadratic, and if each α_k is the value of α that minimizes $F(\underline{x}_k + \alpha\,\underline{d}_k)$, $\alpha > 0$, which is the condition $\underline{d}_k^T\underline{\nabla}F(\underline{x}_k + \alpha_k\underline{d}_k) = 0$, then, in exact arithmetic, the least value of F is calculated after at most n iterations. His arguments include some variable metric points of view, familiar to experts in the theory of relativity, but many researchers including myself do not understand them properly. Therefore other proofs of quadratic termination have been constructed, which depend strongly on the fact that the algorithm with exact line searches gives the conjugacy property $\underline{d}_k^T \nabla^2 F \underline{d}_j = 0$, $j \neq k$, in the quadratic case. Thus the orthogonality conditions

$$\underline{d}_j^T \nabla F(\underline{x}_{k+1}) \;=\; 0, \qquad j = 1, 2, \ldots, k, \tag{2.2}$$

are achieved. There are no restrictions on the choices of $\underline{x}_1$ and the symmetric matrix H_1 for the first iteration, except that H_1 must be positive definite.

The brilliant advantage of this algorithm over classical methods is that it can be applied easily to minimize a general differentiable function F, even if a good initial vector of variables $\underline{x}_1$ is not available, and it gives fast convergence when F is quadratic. Not having to calculate second derivatives is welcome, and it brings two more benefits over the Newton–Raphson procedure. Firstly there is no need to devise a remedy for loss of positive definiteness in $\nabla^2 F(\underline{x})$, and secondly the amount of routine work of each iteration is only $\mathcal{O}(n^2)$ instead of $\mathcal{O}(n^3)$. Another attractive property is invariance under linear transformations of the variables. Specifically, let $\underline{x}_k$, $k = 1, 2, 3, \ldots$, and $\underline{z}_k$, $k = 1, 2, 3, \ldots$, be the vectors of variables that are generated when the algorithm is applied to the functions $F(\underline{x})$, $\underline{x} \in \mathbb{R}^n$, and $F(S^{-1}\underline{z})$, $\underline{z} \in \mathbb{R}^n$, respectively, where S is any constant real $n \times n$ nonsingular matrix. Then, if the initial vector of variables in the second case is $\underline{z}_1 = S\underline{x}_1$, if the initial positive definite matrix is changed from H_1 to $S H_1 S^T$ for the second case, and if there are no changes to the procedure for choosing each step-length α_k, then the second sequence of variables is $\underline{z}_k = S\underline{x}_k$, $k = 1, 2, 3, \ldots$. It follows that good efficiency does not require the variables to be scaled so that their magnitudes are similar. Furthermore, one can simplify the theoretical analysis when F is quadratic by assuming without loss of generality that $\nabla^2 F$ is the unit matrix.

The investigations of Roger Fletcher into Davidon's recent algorithm were similar to my own, so we reported them in a joint paper (Fletcher and Powell, 1963), and the original algorithm has become known as the DFP method. One can view $B_{k+1} = H_{k+1}^{-1}$ as an approximation to $\nabla^2 F$, partly because equation (2.1) gives $H_{k+1}\underline{\gamma}_k = \underline{\delta}_k$, which implies $\underline{\gamma}_k = B_{k+1}\underline{\delta}_k$, while $\underline{\gamma}_k = \nabla^2 F \underline{\delta}_k$ holds when F is quadratic. The matrix $B_{k+1} = H_{k+1}^{-1}$ can be calculated directly from $B_k = H_k^{-1}$, equation (2.1) being equivalent to the formula

$$B_{k+1} \;=\; \left(I - \frac{\underline{\gamma}_k \underline{\delta}_k^T}{\underline{\delta}_k^T \underline{\gamma}_k} \right) B_k \left(I - \frac{\underline{\delta}_k \underline{\gamma}_k^T}{\underline{\delta}_k^T \underline{\gamma}_k} \right) + \frac{\underline{\gamma}_k \underline{\gamma}_k^T}{\underline{\delta}_k^T \underline{\gamma}_k}. \tag{2.3}$$

It is sometimes helpful that working with B_k provides the quadratic model

$$F(\underline{x}_k + \underline{d}) \;\approx\; F(\underline{x}_k) + \underline{d}^T \nabla F(\underline{x}_k) + \tfrac{1}{2} \underline{d}^T B_k \underline{d}, \qquad \underline{d} \in \mathbb{R}^n. \tag{2.4}$$

Expression (2.3) allows the Cholesky factorization of B_{k+1} to be derived from the Cholesky factorization of B_k in $\mathcal{O}(n^2)$ operations. Thus positive definiteness is preserved in the presence of computer rounding errors, and it is inexpensive to obtain the usual search direction $\underline{d}_k = -H_k \nabla F(\underline{x}_k)$ from the linear system

$B_k \underline{d}_k = -\underline{\nabla} F(\underline{x}_k)$. A comparison of equations (2.1) and (2.3) suggests the formula

$$H_{k+1} = \left(I - \frac{\delta_k \gamma_k^T}{\delta_k^T \gamma_k} \right) H_k \left(I - \frac{\gamma_k \delta_k^T}{\delta_k^T \gamma_k} \right) + \frac{\delta_k \delta_k^T}{\delta_k^T \gamma_k}. \tag{2.5}$$

If it replaces equation (2.1) in the DFP method, then we have the well-known BFGS method, which is usually faster than the DFP method in practice.

The other major advance in unconstrained optimization that we consider in this section is the conjugate gradient method of Fletcher and Reeves (1964). It can be applied to general differentiable functions $F(\underline{x})$, $\underline{x} \in \mathbb{R}^n$, it is designed to be efficient when F is quadratic, and it has the strong advantage over the variable metric algorithm of not requiring any $n \times n$ matrices. It can be regarded as an extension of the steepest descent method, retaining $\underline{d}_1 = -\underline{\nabla} F(\underline{x}_1)$, but the search directions of later iterations have the form

$$\underline{d}_k = -\underline{\nabla} F(\underline{x}_k) + \beta_k \underline{d}_{k-1}, \qquad k \geq 2, \tag{2.6}$$

where β_k is allowed to be nonzero. Then a line search picks the step-length that provides the new vector of variables (1.4), which completes the description of the k-th iteration except for the choices of α_k and β_k.

These choices are made in a way that achieves the orthogonality conditions (2.2) for each iteration number k when F is a strictly convex quadratic function, assuming exact arithmetic and termination if $\|\underline{\nabla} F(\underline{x}_k)\| = 0$ occurs. We satisfy $\underline{d}_k^T \underline{\nabla} F(\underline{x}_{k+1}) = 0$ by letting α_k be the α that minimizes $F(\underline{x}_k + \alpha \underline{d}_k)$, $\alpha > 0$, while the $(k-1)$-th of the conditions (2.2) defines β_k. Specifically, because the line search of the previous iteration gives $\underline{d}_{k-1}^T \underline{\nabla} F(\underline{x}_k) = 0$, we require $\underline{d}_{k-1}^T \{ \underline{\nabla} F(\underline{x}_{k+1}) - \underline{\nabla} F(\underline{x}_k) \} = 0$, which is equivalent to $\{ \underline{\nabla} F(\underline{x}_k) - \underline{\nabla} F(\underline{x}_{k-1}) \}^T \underline{d}_k = 0$ in the quadratic case. It follows from equation (2.6) that β_k should take the value

$$\beta_k = \{ \underline{\nabla} F(\underline{x}_k) - \underline{\nabla} F(\underline{x}_{k-1}) \}^T \underline{\nabla} F(\underline{x}_k) \Big/ \{ \underline{\nabla} F(\underline{x}_k) - \underline{\nabla} F(\underline{x}_{k-1}) \}^T \underline{d}_{k-1}$$

$$= \{ \underline{\nabla} F(\underline{x}_k) - \underline{\nabla} F(\underline{x}_{k-1}) \}^T \underline{\nabla} F(\underline{x}_k) \Big/ \|\underline{\nabla} F(\underline{x}_{k-1})\|^2, \tag{2.7}$$

the denominator in the second line being derived from exact line searches and the form of $\underline{d}_{k-1}$. The description of the algorithm is now complete when F is quadratic, and we note that $\nabla^2 F$ is not required. Further analysis in this case can establish the first $k-2$ of the conditions (2.2). It exposes not only the conjugacy property $\underline{d}_k^T \nabla^2 F \underline{d}_j = 0$, $j \neq k$, but also that the gradients $\underline{\nabla} F(\underline{x}_k)$, $k = 1, 2, 3, \ldots$, are mutually orthogonal.

The second line of expression (2.7) states the formula for β_k that is preferred by Polak and Ribière (1969) for general F, but Fletcher and Reeves (1964) propose $\beta_k = \|\underline{\nabla} F(\underline{x}_k)\|^2 / \|\underline{\nabla} F(\underline{x}_{k-1})\|^2$. These two choices are equivalent in the theory of the quadratic case, due to the mutual orthogonality of gradients that has been mentioned, but they are quite different for general F, especially if the changes to $\underline{\nabla} F(\underline{x}_k)$ become relatively small as k is increased. The alternative of Polak and Ribière seems to be more efficient in practice, and it is even better to increase their β_k to zero if it becomes negative. Another reason for modifying

β_k (or α_{k-1}) is that, if β_k is nonzero, then the conditions (1.6) of the previous iteration may fail to supply the descent property $\underline{d}_k^T \nabla F(\underline{x}_k) < 0$.

We see that the conjugate gradient technique is nearly as easy to apply as steepest descents, and usually it provides huge gains in efficiency. The DFP and BFGS algorithms with $H_1 = I$ (the unit matrix) are equivalent to the conjugate gradient method when F is quadratic and all line searches are exact, but the use of the matrices H_k, $k = 1, 2, 3, \ldots$, brings a strong advantage for general F. In order to explain it, we assume that the sequence $\underline{x}_k$, $k = 1, 2, 3, \ldots$, converges to $\underline{x}^*$, say, and that F becomes exactly quadratic only in a neighbourhood of $\underline{x}^*$, which hardly ever happens in practice. The excellent convergence properties of variable metric algorithms are enjoyed automatically when the points $\underline{x}_k$ enter the neighbourhood, without any restrictions on the current $\underline{x}_k$ and the positive definite matrix H_k. On the other hand, the corresponding convergence properties of the conjugate gradient method require a special choice of the initial search direction, $\underline{d}_1 = -\nabla F(\underline{x}_1)$ being suitable, except that the implications of this choice would be damaged by the generality of F on the early iterations. The perfect remedy would set $\beta_k = 0$ as soon as the variables $\underline{x}_k$ stay within the neighbourhood, and perhaps on some earlier iterations too. In practice, β_k can be set to zero when, after the most recent steepest descent iteration, a substantial loss of orthogonality in the sequence of gradients $\nabla F(\underline{x}_k)$ is observed.

3   Unconstrained objective functions for constrained problems

The methods of Section 2 provide huge improvements over classical algorithms for unconstrained optimization. Therefore it was attractive in the 1960s to include constraints on the variables by modifying the objective functions of unconstrained calculations. In particular, the techniques in the book of Fiacco and McCormick (1968) were very popular. Some of them are addressed below.

Let the least value of $F(\underline{x})$, $\underline{x} \in \mathbb{R}^n$, be required, subject to the inequality constraints

$$c_i(\underline{x}) \geq 0, \qquad i = 1, 2, \ldots, m. \tag{3.1}$$

Then a typical objective function of a barrier method has the form

$$\left. \begin{array}{l} \Phi(\underline{x}, \mu) = F(\underline{x}) + \mu \sum_{i=1}^m \{c_i(\underline{x})\}^{-1} \\ \text{or } \Phi(\underline{x}, \mu) = F(\underline{x}) - \mu \sum_{i=1}^m \log\{c_i(\underline{x})\} \end{array} \right\}, \quad \underline{x} \in \mathbb{R}^n, \tag{3.2}$$

if $\underline{x}$ satisfies all the constraints (3.1) as strict inequalities, but otherwise $\Phi(\underline{x}, \mu)$ is defined to be $+\infty$. Here μ is a positive parameter that remains fixed during the unconstrained minimization of $\Phi(\underline{x}, \mu)$, $\underline{x} \in \mathbb{R}^n$. The starting point $\underline{x}_1$ of this calculation has to satisfy $c_i(\underline{x}_1) > 0$, $i = 1, 2, \ldots, m$, because $\Phi(\underline{x}_1, \mu)$ is required to be finite. Let $\underline{x}[\mu]$ be the vector of variables that is produced by this calculation.

The constraints (3.1) are also satisfied as strict inequalities at $\underline{x}[\mu]$, because the unconstrained algorithm provides $\Phi(\underline{x}[\mu], \mu) \leq \Phi(\underline{x}_1, \mu)$ automatically, but it is usual for the solution, $\underline{x}^*$ say, of the original problem to be on the boundary of the feasible region. In this case, the theory of barrier methods requires F and c_i, $i = 1, 2, \ldots, m$, to be continuous functions, and it requires every neighbourhood of $\underline{x}^*$ to include a strictly interior point of the feasible region. Then it is straightforward to establish $F(\underline{x}[\mu]) < F(\underline{x}^*) + \varepsilon$ for sufficiently small μ, where ε is any positive constant, assuming that $\Phi(\underline{x}[\mu], \mu)$ is sufficiently close to the least value of $\Phi(\underline{x}, \mu)$, $\underline{x} \in \mathbb{R}^n$.

Equality constraints, however, cannot be included in barrier function methods, because they cannot be satisfied as strict inequalities. Therefore, when minimizing $F(\underline{x})$, $\underline{x} \in \mathbb{R}^n$, subject to the conditions

$$c_i(\underline{x}) = 0, \qquad i = 1, 2, \ldots, m, \tag{3.3}$$

it was usual to apply an algorithm for unconstrained minimization to the function

$$\left. \begin{array}{l} \Phi(\underline{x}, \mu) = F(\underline{x}) + \mu^{-1} \sum_{i=1}^{m} \{c_i(\underline{x})\}^2 \\ \text{or } \Phi(\underline{x}, \mu) = F(\underline{x}) + \mu^{-1} \sum_{i=1}^{m} |c_i(\underline{x})| \end{array} \right\}, \quad \underline{x} \in \mathbb{R}^n, \tag{3.4}$$

where μ is still a positive parameter, fixed during each unconstrained calculation, that has to become sufficiently small. A new difficulty is shown by the minimization of $F(x) = x^3$, $x \in \mathbb{R}$, subject to $x = 1$, namely that, for any fixed $\mu > 0$, the functions (3.4) are not bounded below. On the other hand, if $\underline{x}[\mu]$ is the minimizer of $\Phi(\underline{x}, \mu)$, $\underline{x} \in \mathbb{R}^n$, if the points $\underline{x}[\mu]$, $\mu > 0$, all lie in a compact region of $\mathbb{R}^n$, and if the objective and constraint functions are continuous, then all limit points of the sequence $\underline{x}[\mu]$ as $\mu \to 0$ are solutions of the original problem. The two main ingredients in a proof of this assertion are that the constraints are satisfied at the limit points, and that, for every positive μ, $F(\underline{x}[\mu])$ is a lower bound on the required value of F.

Penalty function methods are also useful for inequality constraints. If $c_i(\underline{x}) = 0$ were replaced by $c_i(\underline{x}) \geq 0$, then, in expression (3.4), it would be suitable to replace the terms $\{c_i(\underline{x})\}^2$ and $|c_i(\underline{x})|$ by $\{\min[0, c_i(\underline{x})]\}^2$ and $\max[0, -c_i(\underline{x})]$, respectively.

The dependence of the error $\underline{x}[\mu] - \underline{x}^*$ on μ, for both the inequality and equality constrained problems that have been mentioned, can be investigated by comparing the condition for an unconstrained minimum of $\Phi(\underline{x}, \mu)$, $\underline{x} \in \mathbb{R}^n$, with the KKT conditions for a solution of the original problem. We consider this approach briefly when the constraints are the equations (3.3), when Φ is the first of the functions (3.4), when the objective and constraint functions have continuous first derivatives, when $\underline{\nabla} F(\underline{x}^*)$ is nonzero, and when the constraint gradients $\underline{\nabla} c_i(\underline{x}^*)$, $i = 1, 2, \ldots, m$, are linearly independent. Then $\underline{\nabla} \Phi(\underline{x}[\mu], \mu) = 0$ is the equation

$$\underline{\nabla} F(\underline{x}[\mu]) + 2\mu^{-1} \sum_{i=1}^{m} c_i(\underline{x}[\mu]) \underline{\nabla} c_i(\underline{x}[\mu]) = 0, \tag{3.5}$$

while the first order KKT conditions include the existence of unique Lagrange multipliers $\lambda_i^* \in \mathbb{R}$, $i = 1, 2, \ldots, m$, not all zero, such that $\underline{\nabla} F(\underline{x}^*)$ can be expressed in the form

$$\underline{\nabla} F(\underline{x}^*) = \sum_{i=1}^{m} \lambda_i^* \underline{\nabla} c_i(\underline{x}^*). \tag{3.6}$$

Therefore, if $\underline{x}[\mu]$ tends to $\underline{x}^*$ as expected when $\mu \to 0$, we have the estimates $c_i(\underline{x}[\mu]) \approx -\frac{1}{2}\mu\lambda_i^*$, $i = 1, 2, \ldots, m$. It follows that the distance from $\underline{x}[\mu]$ to any point in $\mathbb{R}^n$ that satisfies the constraints is at least of magnitude μ. Typically, $\|\underline{x}[\mu] - \underline{x}^*\|$ is also of this magnitude, but there are exceptions, such as the minimization of $x_1^4 + x_1 x_2 + x_2$, $\underline{x} \in \mathbb{R}^2$, subject to $x_2 = 0$. We will return to this example later.

The efficiency of these barrier and penalty function methods depends strongly on suitable stopping conditions for the unconstrained calculations, on the size of the reductions in μ, and on obtaining a good starting vector and second derivative estimates for each new unconstrained problem from the sequence of unconstrained problems that have been solved already. Research on interior point methods has given much attention to these questions during the last twenty years, because the path $\underline{x}[\mu]$, $\mu > 0$, in $\mathbb{R}^n$ is a part of the central path of a primal-dual algorithm (see Wright, 1997, for instance). In the early 1970s, however, barrier and penalty function methods became unpopular, due to the development of new techniques for constraints that avoid the difficulties that arise when μ is tiny. In particular, the functions $\Phi(\underline{x}, \mu)$, $\underline{x} \in \mathbb{R}^n$, tend to have some huge first derivatives, so a descent method for unconstrained minimization can reach the bottom of a cliff easily. Then the remainder of the route to $\underline{x}[\mu]$ has to stay at the bottom of the cliffs that are caused by the barrier or penalty terms, which is a daunting situation, especially if the constraints are nonlinear.

The augmented Lagrangian method, proposed by Hestenes (1969) and Powell (1969) independently, is a highly useful extension to the minimization of the first of the functions (3.4), when seeking the least value of $F(\underline{x})$, $\underline{x} \in \mathbb{R}^n$, subject to the equality constraints (3.3). The new penalty function has the form

$$\Lambda(\underline{x}, \underline{\lambda}, \mu) = F(\underline{x}) - \sum_{i=1}^{m} \lambda_i \, c_i(\underline{x}) + \mu^{-1} \sum_{i=1}^{m} \{c_i(\underline{x})\}^2, \qquad \underline{x} \in \mathbb{R}^n, \quad (3.7)$$

its unconstrained minimum being calculated approximately for each fixed choice of the parameters $\underline{\lambda} \in \mathbb{R}^m$ and $\mu > 0$. Let this calculation give the vector of variables $\underline{x}[\underline{\lambda}, \mu]$. The main feature of the augmented Lagrangian method is that it tries to satisfy the constraints $c_i(\underline{x}[\underline{\lambda}, \mu]) = 0$, $i = 1, 2, \ldots, m$, by adjusting $\underline{\lambda} \in \mathbb{R}^m$ without further reductions in μ when μ becomes sufficiently small. It follows from equation (3.6) and from the assumed linear independence of the constraint gradients that, if $\mu > 0$, then the solution $\underline{x}^*$ of the original problem is at the unconstrained minimum of the function (3.7) only if $\underline{\lambda}$ has the components $\lambda_i = \lambda_i^*$, $i = 1, 2, \ldots, m$.

In the awkward problem that has been mentioned of minimizing $x_1^4 + x_1 x_2 + x_2$, $\underline{x} \in \mathbb{R}^2$, subject to $x_2 = 0$, we find $\lambda_1^* = 1$ and that expression (3.7) is the function

$$\Lambda(\underline{x}, \underline{\lambda}^*, \mu) = x_1^4 + x_1 x_2 + \mu^{-1} x_2^2, \qquad \underline{x} \in \mathbb{R}^2, \quad (3.8)$$

which is stationary at the required solution $\underline{x}^* = 0$. Unfortunately this stationary point is never a minimum when μ is fixed and positive. However, the second order condition $\underline{d}^T \{\nabla^2 F(\underline{x}^*) - \sum_{i=1}^{m} \lambda_i^* \nabla^2 c_i(\underline{x}^*)\} \underline{d} > 0$, where $\underline{d}$ is any nonzero vector that is orthogonal to $\underline{\nabla} c_i(\underline{x}^*)$, $i = 1, 2, \ldots, m$, is usually satisfied for the given problem. In this case, the function (3.7) with $\underline{\lambda} = \underline{\lambda}^*$ is not only stationary at $\underline{x}=\underline{x}^*$, but also the second derivative matrix $\nabla^2 \Lambda(\underline{x}^*, \underline{\lambda}^*, \mu)$ is positive definite for sufficiently small μ. It follows that $\underline{x}^*$ can be calculated by the unconstrained minimization of $\Lambda(\underline{x}, \underline{\lambda}^*, \mu)$, $\underline{x} \in \mathbb{R}^n$.

The initial choice of μ and any later reductions should provide suitable local minima in the unconstrained calculations and should help the achievement of $\underline{\lambda} \rightarrow \underline{\lambda}^*$. Usually the components of $\underline{\lambda}$ are set to zero initially. A convenient way of adjusting $\underline{\lambda}$ is based on the remark that, if $\underline{x}$ is a stationary point of the function (3.7), then it satisfies the equation

$$\nabla \Lambda(\underline{x}, \underline{\lambda}, \mu) = \nabla F(\underline{x}) - \sum_{i=1}^m \{\lambda_i - 2\mu^{-1} c_i(\underline{x})\} \nabla c_i(\underline{x}) = 0. \tag{3.9}$$

Specifically, a comparison of equations (3.6) and (3.9) suggests the formula

$$\lambda_i \leftarrow \lambda_i - 2\mu^{-1} c_i(\underline{x}[\underline{\lambda}, \mu]), \qquad i=1, 2, \ldots, m, \tag{3.10}$$

where "$\leftarrow$" denotes "is replaced by". The success of this technique requires μ to be sufficiently small. Other techniques for updating $\underline{\lambda}$ have been derived from the remark that $\underline{\lambda}^*$ should be the value of $\underline{\lambda}$ that maximizes $\Lambda(\underline{x}[\underline{\lambda}, \mu], \underline{\lambda}, \mu)$, $\underline{\lambda} \in \mathbb{R}^m$. Indeed, the calculation of $\underline{x}[\underline{\lambda}, \mu]$ should provide the bound

$$\Lambda(\underline{x}[\underline{\lambda}, \mu], \underline{\lambda}, \mu) \leq \Lambda(\underline{x}^*, \underline{\lambda}, \mu) = F(\underline{x}^*) = \Lambda(\underline{x}^*, \underline{\lambda}^*, \mu) \tag{3.11}$$

for every choice of $\underline{\lambda}$, the last two equations being elementary consequences of the constraints $c_i(\underline{x}^*)=0$, $i=1, 2, \ldots, m$.

The augmented Lagrangian method became even more useful when Rockafellar (1973) proposed and analysed a version of expression (3.7) that is suitable for inequality constraints. Specifically, when the original problem is the minimization of $F(\underline{x})$, $\underline{x} \in \mathbb{R}^n$, subject to the conditions (3.1), then $\underline{x}[\underline{\lambda}, \mu]$ is calculated by applying an algorithm for unconstrained minimization to the function

$$\Lambda(\underline{x}, \underline{\lambda}, \mu) = F(\underline{x}) + \mu^{-1} \sum_{i=1}^m \{\min[0, c_i(\underline{x}) - \tfrac{1}{2}\mu\lambda_i]\}^2, \qquad \underline{x} \in \mathbb{R}^n, \tag{3.12}$$

for a sequence of fixed values of $\underline{\lambda} \in \mathbb{R}^m$ and $\mu \in \mathbb{R}$. Again the constraints are satisfied by adjusting only $\underline{\lambda}$ if possible for sufficiently small μ. We see that, if $\underline{x}$ is a stationary point of the new $\Lambda(\underline{x}, \underline{\lambda}, \mu)$, $\underline{x} \in \mathbb{R}^n$, then it satisfies the equation

$$\nabla F(\underline{x}) - \sum_{i=1}^m \max[0, \lambda_i - 2\mu^{-1} c_i(\underline{x})] \nabla c_i(\underline{x}) = 0. \tag{3.13}$$

Therefore we modify formula (3.10) for adjusting $\underline{\lambda}$ by letting $\max[0, \lambda_i - 2\mu^{-1} c_i(\underline{x})]$ at $\underline{x} = \underline{x}[\underline{\lambda}, \mu]$ be the new right hand side. Thus the components of $\underline{\lambda}$ are non-negative, as required in the KKT condition (3.6) of the original problem when the constraints are inequalities. Further, λ_i^* should be zero in equation (3.6) for every i that satisfies $c_i(\underline{x}^*)>0$, and, if μ is sufficiently small, the modification of formula (3.10) gives λ_i this property automatically. A mixture of equality and inequality constraints can be treated by taking their contributions to $\Lambda(\underline{x}, \underline{\lambda}, \mu)$ from expressions (3.7) and (3.12), respectively.

4   Sequential quadratic programming

Often the methods of the last section are too elaborate and too sophisticated. An extreme example is the minimization of $F(\underline{x})$, $\underline{x} \in \mathbb{R}^n$, subject to $x_n = 0$. The constraint allows the number of variables to be decreased by one, and then a single unconstrained calculation with $n-1$ variables can be solved, instead of a sequence of unconstrained calculations with n variables. The sequence of subproblems can also be avoided when the constraints are nonlinear by making linear approximations to the constraints. In particular, if the least value of $F(\underline{x})$ is required subject to the equality constraints (3.3), and if the objective and constraint functions have continuous second derivatives, then one can apply Newton's method for solving nonlinear equations to the system that is given by the first order KKT conditions at the solution. That approach has several disadvantages. Many of them were removed by the development of sequential quadratic programming (SQP), which is addressed below, because SQP became a popular successor to the augmented Lagrangian method for constrained calculations in the late 1970s.

In the application of Newton's method that has just been mentioned, the unknowns are not only the variables x_i, $i = 1, 2, \ldots, n$, but also the Lagrange multipliers of condition (3.6). Specifically, we seek vectors $\underline{x} \in \mathbb{R}^n$ and $\underline{\lambda} \in \mathbb{R}^m$ that satisfy the square system of equations

$$\left.\begin{array}{l} \underline{\nabla} F(\underline{x}) - \sum_{i=1}^{m} \lambda_i \underline{\nabla} c_i(\underline{x}) = 0 \quad \text{and} \\ -c_i(\underline{x}) = 0, \qquad i = 1, 2, \ldots, m, \end{array}\right\} \tag{4.1}$$

the signs of the equality constraints (3.3) being reversed in order that the Jacobian matrix of Newton's method is symmetric. Let $\underline{f}(\underline{x}, \underline{\lambda})$, $\underline{x} \in \mathbb{R}^n$, $\underline{\lambda} \in \mathbb{R}^m$, be the vector in $\mathbb{R}^{m+n}$ whose components are the left hand sides of expression (4.1). As in Section 1, the k-th iteration of Newton's method without line searches calculates $\underline{x}_{k+1} = \underline{x}_k + \underline{d}_k$ and $\underline{\lambda}_{k+1} = \underline{\lambda}_k + \underline{\eta}_k$ by equating to zero a first order Taylor series approximation to the function $\underline{f}(\underline{x}_k + \underline{d}, \underline{\lambda}_k + \underline{\eta})$, $\underline{d} \in \mathbb{R}^n$, $\underline{\eta} \in \mathbb{R}^m$. Specifically, the analogue of equation (1.3) is that $\underline{d}_k$ and $\underline{\eta}_k$ are derived from the linear system

$$\left(\begin{array}{c|c} W(\underline{x}_k, \underline{\lambda}_k) & -J(\underline{x}_k)^T \\ \hline -J(\underline{x}_k) & 0 \end{array}\right) \left(\begin{array}{c} \underline{d}_k \\ \underline{\eta}_k \end{array}\right) = \left(\begin{array}{c} -\underline{\nabla} F(\underline{x}_k) + J(\underline{x}_k)^T \underline{\lambda}_k \\ \underline{c}(\underline{x}_k) \end{array}\right), \tag{4.2}$$

where $W(\underline{x}, \underline{\lambda}) = \nabla^2 F(\underline{x}) - \sum_{i=1}^{m} \lambda_i \nabla^2 c_i(\underline{x})$, where $J(\underline{x})$ is now the $m \times n$ matrix that has the elements

$$[J(\underline{x})]_{ij} = dc_i(\underline{x})/dx_j, \quad 1 \le i \le m, \quad 1 \le j \le n, \tag{4.3}$$

and where $\underline{c}(\underline{x})$ is the vector in $\mathbb{R}^m$ with the components $c_i(\underline{x})$, $i = 1, 2, \ldots, m$.

This application of Newton's method has the following three disadvantages. The calculation breaks down if the partitioned matrix of the linear system (4.2) becomes singular. No attempt is made to help convergence when good initial values of the variables are not available. The minimization ingredient of the original problem is absent from the formulation (4.1). On the other hand, the method provides a highly useful answer to a very important question, which is to identify the second derivatives that are usually sufficient for a fast rate of convergence. We see that the k-th iteration in the previous paragraph requires second derivatives of the objective and constraint functions only to assemble the matrix $W(\underline{x}_k, \underline{\lambda}_k)$. Therefore, when second derivatives are estimated, one should construct an approximation to the combination $\nabla^2 F(\underline{x}) - \sum_{i=1}^{m} \lambda_i \nabla^2 c_i(\underline{x})$, which is much more convenient than estimating all the matrices $\nabla^2 F(\underline{x})$ and $\nabla^2 c_i(\underline{x})$, $i = 1, 2, \ldots, m$, separately.

We recall from Section 2 that variable metric algorithms for unconstrained optimization bring huge advantages over the Newton–Raphson method by working with positive definite approximations to $\nabla^2 F$. Similar gains can be achieved in constrained calculations over the Newton iteration above by making a positive definite approximation to $W(\underline{x}_k, \underline{\lambda}_k)$ in the system (4.2). We let B_k be such an approximation, and we consider the minimization of the strictly convex quadratic function

$$Q_k(\underline{x}_k + \underline{d}) = F(\underline{x}_k) + \underline{d}^T \nabla F(\underline{x}_k) + \tfrac{1}{2} \underline{d}^T B_k \underline{d}, \qquad \underline{d} \in \mathbb{R}^n, \tag{4.4}$$

subject to the linear constraints

$$c_i(\underline{x}_k) + \underline{d}^T \nabla c_i(\underline{x}_k) = 0, \qquad i = 1, 2, \ldots, m, \tag{4.5}$$

still assuming that the constraint gradients are linearly independent. The vector $\underline{d} = \underline{d}_k$ is the solution to this problem if and only if it satisfies the constraints (4.5) and the gradient $\nabla Q_k(\underline{x}_k + \underline{d}_k) = \nabla F(\underline{x}_k) + B_k \underline{d}_k$ is in the linear space spanned by $\nabla c_i(\underline{x}_k)$, $i = 1, 2, \ldots, m$. In other words, $\underline{d}_k$ has to satisfy the equations (4.2) with $W(\underline{x}_k, \underline{\lambda}_k)$ replaced by B_k, the partitioned matrix of the new system being nonsingular due to the given assumptions. Thus the calculation of $\underline{d}_k$ by the Newton iteration is equivalent to the solution of the strictly convex quadratic programming problem, which captures the minimization ingredient that has been mentioned. A more important benefit of the alternative calculation of $\underline{d}_k$ is that it has a natural extension for inequality constraints, by continuing to let $\underline{d}_k$ be the vector $\underline{d}$ that minimizes the strictly convex quadratic function (4.4) subject to first order Taylor series approximations to all the constraints. Specifically, for each constraint index i, the original constraint $c_i(\underline{x}) = 0$ or $c_i(\underline{x}) \geq 0$ contributes the condition $c_i(\underline{x}_k) + \underline{d}^T \nabla c_i(\underline{x}_k) = 0$ or $c_i(\underline{x}_k) + \underline{d}^T \nabla c_i(\underline{x}_k) \geq 0$, respectively, to the quadratic programming problem, without any change to $Q_k(\underline{x}_k + \underline{d})$, $\underline{d} \in \mathbb{R}^n$, after B_k has been chosen.

The DFP formula (2.3) (or the well-known BFGS formula) may be used to define B_{k+1} for the next iteration, where $\underline{\delta}_k$ is the step $\underline{x}_{k+1} - \underline{x}_k$ as before, but the selection of $\underline{\gamma}_k$ requires further consideration. The updating formula gives $B_{k+1} \underline{\delta}_k = \underline{\gamma}_k$, so $\underline{\gamma}_k$ must satisfy $\underline{\delta}_k^T \underline{\gamma}_k > 0$, in order that B_{k+1} inherits positive definiteness from B_k. On the other hand, because B_{k+1} should be an estimate of the combination $\nabla^2 F(\underline{x}_{k+1}) - \sum_{i=1}^{m} \lambda_i \nabla^2 c_i(\underline{x}_{k+1})$, as mentioned already, it it suitable to let the difference

$$\widehat{\gamma}_k = \nabla F(\underline{x}_{k+1}) - \nabla F(\underline{x}_k) - \sum_{i=1}^{m} \lambda_i \{\nabla c_i(\underline{x}_{k+1}) - \nabla c_i(\underline{x}_k)\} \qquad (4.6)$$

be a provisional choice of γ_k, where the multipliers λ_i, $i = 1, 2, \ldots, m$, can be taken from the quadratic programming problem that defines $\underline{d}_k$, even if some of the constraints are inequalities. It is possible, however, for the original problem to be the minimization of $F(\underline{x}) = -\frac{1}{2}\|\underline{x}\|^2$, $\underline{x} \in \mathbb{R}^n$, subject to constraints that are all linear. Then equation (4.6) gives $\widehat{\gamma}_k = -\underline{x}_{k+1} + \underline{x}_k = -\delta_k$, which implies $\delta_k^T \widehat{\gamma}_k < 0$, although we require $\delta_k^T \gamma_k > 0$. Therefore the form $\gamma_k = \theta_k \widehat{\gamma}_k + (1 - \theta_k) B_k \delta_k$ is proposed in Powell (1978) for the DFP or BFGS updating formula, where θ_k is the largest number from $[0, 1]$ that satisfies $\delta_k^T \gamma_k \geq 0.1 \delta_k^T B_k \delta_k$. A device of this kind was necessary in order to provide software.

Another challenge for SQP software is forcing convergence from poor starting points. A remedy in Section 1 is to seek $\underline{x}_{k+1}$ by a line search from $\underline{x}_k$ along the direction $\underline{d}_k$, but, if all the early iterations require tiny step-lengths, then the progress towards constraint boundaries is very slow, even if the constraints are linear. Therefore some implementations of the SQP method employ two kinds of changes to the variables, namely horizontal and vertical steps, where horizontal steps include line searches and try to reduce the objective function without worsening constraint violations, and where the main purpose of vertical steps is to correct the departures from feasibility (see Coleman and Conn, 1982, for instance). Several techniques have also been proposed for deciding whether or not to accept a trial step in a line search, the difficulty being that improvements in the objective function and decreases in constraint violations may not occur together. The usual compromise is to seek a reduction in the penalty function

$$\Phi(\underline{x}, \mu) = F(\underline{x}) + \mu^{-1}\{ \sum_{i \in \mathcal{E}} |c_i(\underline{x})| + \sum_{i \in \mathcal{I}} \max[0, -c_i(\underline{x})] \}, \qquad (4.7)$$

where $\mathcal{E}$ and $\mathcal{I}$ contain the indices of the equality and inequality constraints, respectively, and where μ has to be selected automatically. Alternatively, instead of taking dubious decisions in the line searches, one can keep options open by applying the filter method of Fletcher and Leyffer (2002). Many different versions of the SQP method have been developed for constrained calculations when first derivatives are available, and usually they are excellent at keeping down the total number of function and gradient evaluations.

5   Trust region methods

We recall that, in line search methods for forcing convergence from general starting points, the sequence of iterations gives the variables

$$\underline{x}_{k+1} = \underline{x}_k + \alpha_k \underline{d}_k, \qquad k = 1, 2, 3, \ldots, \qquad (5.1)$$

where usually the search direction $\underline{d}_k$ is derived from a simple model of the original problem, and where the choice of the step-length α_k should make $\underline{x}_{k+1}$ better than $\underline{x}_k$ according to the criteria of the original problem, the simplest example being the condition $F(\underline{x}_{k+1}) < F(\underline{x}_k)$ when the least value of $F(\underline{x})$, $\underline{x} \in \mathbb{R}^n$, is required. We expect the model of the k-th iteration to provide useful accuracy in a neighbourhood of $\underline{x}_k$, but $\underline{x}_k + \underline{d}_k$ may be far from that neighbourhood, so often the step-lengths of line search methods are substantially less than one for many consecutive iterations. Then it is reasonable to take the view that each new value of $\|\underline{x}_{k+1} - \underline{x}_k\|$ is not going to be much larger than the magnitudes of the changes to the variables of recent iterations. Under this assumption, one may be able to make much better use of the simple model. For example, moves to constraint boundaries can be made more quickly in the situation that is mentioned in the last paragraph of Section 4. Therefore a bound of the form $\|\underline{d}_k\| \le \Delta_k$ is imposed by a trust region method, the remaining freedom in $\underline{d}_k$ being taken up by consideration of the current simple model. The positive parameter Δ_k is chosen automatically before the start of the k-th iteration. Some details and advantages of this technique are addressed below, because, since the 1970s, trust region methods have become fundamental within many highly successful algorithms for optimization.

We begin with the unconstrained minimization of $F(\underline{x})$, $\underline{x} \in \mathbb{R}^n$, when first derivatives are available, and when the calculation of $\underline{x}_{k+1}$ from $\underline{x}_k$ employs the model

$$F(\underline{x}_k + \underline{d}) \approx Q_k(\underline{x}_k + \underline{d}) = F(\underline{x}_k) + \underline{d}^T \underline{\nabla} F(\underline{x}_k) + \tfrac{1}{2} \underline{d}^T B_k \underline{d}, \qquad \underline{d} \in \mathbb{R}^n, \quad (5.2)$$

as in expressions (2.4) and (4.4), but now there is no need for the symmetric matrix B_k to be positive definite. We assume that termination occurs if $\|\underline{\nabla} F(\underline{x}_k)\|$ is sufficiently small. Otherwise, we require $\underline{d}_k$ to be an estimate of the vector $\underline{d}$ that minimizes $Q_k(\underline{x}_k + \underline{d})$, $\underline{d} \in \mathbb{R}^n$, subject to $\|\underline{d}\| \le \Delta_k$. If $\underline{d}_k$ is an exact solution to this subproblem, then there exists $\lambda_k \ge 0$ such that the equation

$$(B_k + \lambda_k I)\, \underline{d}_k = -\underline{\nabla} F(\underline{x}_k) \tag{5.3}$$

holds, and $B_k + \lambda_k I$ is positive definite or semi-definite, where I is the identity matrix. Thus reliable procedures for calculating $\underline{d}_k$ with control of accuracy are given by Moré and Sorensen but often they are too expensive when n is large. Instead it is usual to apply the conjugate gradient minimization procedure of Section 2 to the quadratic model (5.2), starting at $\underline{d} = 0$. It generates a piecewise linear path in $\mathbb{R}^n$, the difference between the end and the beginning of the ℓ-th line segment of the path being the change that is made to the vector of variables $\underline{d}$ on the ℓ-th iteration. The conjugate gradient iterations are terminated if the path reaches the boundary of the region $\{\underline{d} : \|\underline{d}\| \le \Delta_k\}$, or if the reduction in $Q(\underline{x}_k + \underline{d})$ by an iteration is much less than the total reduction so far. Then $\underline{d}_k$ is chosen to be the final point of the path, except that some algorithms seek further reductions in $Q_k(\underline{x}_k + \underline{d})$ in the case $\|\underline{d}_k\| = \Delta_k$ (Conn, Gould and Toint, 2000).

After picking $\underline{d}_k$, the new function value $F(\underline{x}_k + \underline{d}_k)$ is calculated. The ratio

$$\rho_k = \{F(\underline{x}_k) - F(\underline{x}_k + \underline{d}_k)\} / \{Q_k(\underline{x}_k) - Q_k(\underline{x}_k + \underline{d}_k)\} \tag{5.4}$$

is important, because a value close to one suggests that the current model is good for predicting the behaviour of $F(\underline{x}_k+\underline{d})$, $\|\underline{d}\| \leq \Delta_k$. Therefore the value of Δ_{k+1} for the next iteration may be set to $\max[\Delta_k, 2\|\underline{d}_k\|]$, Δ_k or $\frac{1}{2}\|\underline{d}_k\|$ in the cases $\rho_k \geq 0.8$, $0.2 \leq \rho_k < 0.8$ or $\rho_k < 0.2$, respectively, for example. No other values of F are calculated on the k-th iteration of most trust region methods, $\underline{x}_{k+1}$ being either $\underline{x}_k$ or $\underline{x}_k+\underline{d}_k$. It seems obvious to prefer $\underline{x}_{k+1} = \underline{x}_k + \underline{d}_k$ whenever the strict reduction $F(\underline{x}_k + \underline{d}_k) < F(\underline{x}_k)$ is achieved, which is the condition $\rho_k > 0$. Many trust region algorithms, however, set $\underline{x}_{k+1}$ to $\underline{x}_k + \underline{d}_k$ only if ρ_k is sufficiently large. If $F(\underline{x}_k+\underline{d}_k) \geq F(\underline{x}_k)$ occurs in a trust region method, then the conditions $\underline{x}_{k+1} = \underline{x}_k$ and $\|\underline{d}_{k+1}\| \leq \Delta_{k+1} < \|\underline{d}_k\|$ are satisfied. Hence, if the vector $\underline{x}_{k+1}+\underline{d}_{k+1}$ of the $(k+1)$-th iteration is regarded as the result of a step from $\underline{x}_k$, then the length of the step is less than $\|\underline{d}_k\|$ automatically. Thus trust region methods include a main ingredient of line search methods. Attention is given later to the choice of the new matrix B_{k+1} at the end of the k-th iteration.

As in Section 4, a difficulty in constrained calculations is the need for a balance between reducing $F(\underline{x})$, $\underline{x} \in \mathbb{R}^n$, and correcting violations of the constraints. We retain the compromise of the penalty function (4.7), and we estimate $\Phi(\underline{x}_k+\underline{d},\mu)$, $\underline{d}\in\mathbb{R}^n$, by the model

$$\Xi_k(\underline{x}_k+\underline{d},\mu) = F(\underline{x}_k) + \underline{d}^T\nabla F(\underline{x}_k) + \tfrac{1}{2}\underline{d}^T B_k \underline{d} + \mu^{-1}\{\textstyle\sum_{i\in\mathcal{E}} |c_i(\underline{x}_k) + \underline{d}^T\nabla c_i(\underline{x}_k)|$$

$$+ \textstyle\sum_{i\in\mathcal{I}} \max[0, -c_i(\underline{x}_k) - \underline{d}^T\nabla c_i(\underline{x}_k)]\}, \qquad \underline{d}\in\mathbb{R}^n, \qquad (5.5)$$

which reduces to expression (5.2) if there are no constraints. It is usual to terminate the sequence of iterations if the residuals of the first order KKT conditions are sufficiently small at $\underline{x} = \underline{x}_k$. Otherwise, $\underline{d}_k$ and μ are chosen in a way that satisfies $\|\underline{d}_k\| \leq \Delta_k$ and $\Xi_k(\underline{x}_k+\underline{d}_k,\mu) < \Xi_k(\underline{x}_k,\mu)$. Let $F(\cdot)$ and $Q_k(\cdot)$ be replaced by $\Phi(\cdot,\mu)$ and $\Xi_k(\cdot,\mu)$ throughout the remarks of the previous paragraph, the new version of the definition (5.4) being the ratio

$$\rho_k = \{\Phi(\underline{x}_k,\mu) - \Phi(\underline{x}_k+\underline{d}_k,\mu)\} / \{\Xi_k(\underline{x}_k,\mu) - \Xi_k(\underline{x}_k+\underline{d}_k,\mu)\}. \qquad (5.6)$$

The modified remarks give suitable techniques for choosing Δ_{k+1} and $\underline{x}_{k+1}$ in calculations with constraints on the variables.

If the ∞-norm is used instead of the 2-norm in the bound $\|\underline{d}\| \leq \Delta_k$, then the minimization of the function (5.5) for fixed μ subject to the bound is a quadratic programming problem. Thus the $\underline{d}_k$ of the previous paragraph can be calculated (Fletcher, 1985), with occasional decreases in μ if necessary in order to give enough weight to the constraints. Another way of generating $\underline{d}_k$ begins by letting $\widehat{\underline{d}}_k$ be an estimate of the $\underline{d}$ that minimizes $\Gamma_k(\underline{d})$, $\underline{d}\in\mathbb{R}^n$, subject to $\|\underline{d}\| \leq \frac{1}{2}\Delta_k$, where $\Gamma_k(\underline{d})$ is the term inside the braces of expression (5.5). Then $\underline{d}_k$ has to satisfy $\|\underline{d}_k\| \leq \Delta_k$ and $\Gamma_k(\underline{d}_k) \leq \Gamma_k(\widehat{\underline{d}}_k)$, which leaves some freedom in $\underline{d}_k$. It is taken up by trying to make $Q_k(\underline{x}_k+\underline{d}_k)$ substantially smaller than $Q_k(\underline{x}_k+\widehat{\underline{d}}_k)$, where $Q_k(\underline{x}_k+\underline{d})$ is still the quadratic term (5.2). This technique has the property that $\underline{d}_k$ is independent of μ, which is adjusted separately in a way that controls the required reduction $\Xi_k(\underline{x}_k+\underline{d}_k,\mu) < \Xi_k(\underline{x}_k,\mu)$.

Several advantages are provided by the fact that, in trust region methods, the second derivative matrix B_k of the model does not have to be positive definite. In particular, if the sparsity structure of $\nabla^2 F$ is known in unconstrained optimization, then B_{k+1} may be required to have the same structure in addition to satisfying the equation $B_{k+1}\underline{\delta}_k = \underline{\gamma}_k$ of Section 2, which may not allow B_{k+1} to be positive definite, even if we retain $\underline{\delta}_k^T \gamma_k > 0$. Moreover, we recall from Section 4 that, in constrained calculations, it is suitable to replace the condition $B_{k+1}\underline{\delta}_k = \underline{\gamma}_k$ by $B_{k+1}\underline{\delta}_k = \underline{\hat{\gamma}}_k$, where $\underline{\hat{\gamma}}_k$ is the difference (4.6). There is now no need for an unwelcome device to maintain positive definiteness, as described after equation (4.6). In both of these situations the conditions on the elements of B_{k+1} are linear equality constraints. A highly successful and convenient way of taking up the freedom in B_{k+1} is to minimize $\|B_{k+1}-B_k\|_F$, where the subscript F denotes the Frobenius norm. In other words, we let the new model be as close as possible to the old model subject to the linear constraints, where closeness is measured by the sum of squares of the changes to the elements of the second derivative matrix of the model. Some very useful properties of this technique are given in the next section.

Trust region methods are also more robust than line search methods when the Newton iteration (1.3) is modified, in case the starting point $\underline{x}_1$ is not "sufficiently close" to a solution. We recall that a line search method applies formula (1.4), but a trust region method would choose between the alternatives $\underline{x}_{k+1} = \underline{x}_k + \underline{d}_k$ and $\underline{x}_{k+1} = \underline{x}_k$, where $\underline{d}_k$ is an estimate of the vector $\underline{d}$ that minimizes $\|\underline{f}(\underline{x}_k) + J(\underline{x}_k)\underline{d}\|$ subject to $\|\underline{d}\| \leq \Delta_k$. The usual ways of selecting $\underline{x}_{k+1}$ and Δ_{k+1} for the next iteration are similar to those that have been described already.

6   Further remarks

In my experience, the question that has been most useful to the development of successful algorithms for unconstrained optimization is "Does the method work well when the objective function is quadratic?". The answer is very welcome and encouraging for the updating of second derivative matrices of quadratic models by the symmetric Broyden method, which is the technique of taking up freedom in the new model by minimizing $\|B_{k+1} - B_k\|_F$, mentioned in the paragraph before last. We are going to consider this method in unconstrained calculations when the current quadratic model has the form

$$F(\underline{x}_k + \underline{d}) \approx Q_k(\underline{x}_k + \underline{d}) = F(\underline{x}_k) + \underline{d}^T \underline{g}_k + \tfrac{1}{2}\underline{d}^T B_k \underline{d}, \qquad \underline{d} \in \mathbb{R}^n, \qquad (6.1)$$

where $F(\underline{x}_k)$ and B_k are retained from expression (5.2), but $\underline{g}_k$ is allowed to be an estimate of $\underline{\nabla} F(\underline{x}_k)$ that is given to the k-th iteration, which is useful if first derivatives of F are not available.

Some constraints on the parameters of the new model

$$Q_{k+1}(\underline{x}_{k+1} + \underline{d}) = F(\underline{x}_{k+1}) + \underline{d}^T \underline{g}_{k+1} + \tfrac{1}{2}\underline{d}^T B_{k+1}\underline{d}, \qquad \underline{d} \in \mathbb{R}^n, \qquad (6.2)$$

have been stated already for algorithms that employ first derivatives. In addition to $g_j = \nabla F(x_j)$, $j = 1, 2, 3, \ldots$, they include the equation

$$B_{k+1} \underline{\delta}_k = \underline{\gamma}_k = \nabla F(x_k + \underline{\delta}_k) - \nabla F(x_k), \qquad (6.3)$$

where $\underline{\delta}_k$ is $x_{k+1} - x_k$ or $\underline{d}_k$ in a line search or trust region method, respectively. In algorithms without derivatives, however, the new model Q_{k+1} may be derived from the current model Q_k and from interpolation conditions of the form

$$Q_{k+1}(\underline{z}_j) = F(\underline{z}_j), \qquad j = 1, 2, \ldots, m, \qquad (6.4)$$

where the points $\underline{z}_j$, $j = 1, 2, \ldots, m$, are chosen automatically, one of them being x_{k+1}. I prefer to keep m fixed at about $2n + 1$ and to change only one of the interpolation points on each iteration, which can provide suitable data for the selection of both g_{k+1} and B_{k+1}. The matrix B_{k+1} is required to be symmetric in all of these algorithms, and sometimes B_{k+1} is given the sparsity structure of $\nabla^2 F$.

Let $F(x)$, $x \in \mathbb{R}^n$, be a quadratic function. Then all the constraints on the parameters of Q_{k+1} in the previous paragraph are satisfied if we pick $Q_{k+1} \equiv F$. It follows from the linearity of the constraints that they allow any multiple of the difference $F - Q_{k+1}$ to be added to Q_{k+1}. Therefore, if B_{k+1} is calculated by minimizing $\|B_{k+1} - B_k\|_F$ subject to the constraints, which is the symmetric Broyden method, then the least value of $\phi(\theta) = \|B_{k+1} - B_k + \theta (\nabla^2 F - B_{k+1})\|_F^2$, $\theta \in \mathbb{R}$, occurs at $\theta = 0$. We consider this remark algebraically by introducing the notation $\langle V, W \rangle$ for the sum $\sum_{i=1}^n \sum_{j=1}^n V_{ij} W_{ij}$, where V and W are any $n \times n$ symmetric matrices. The definition of the Frobenius norm gives the expression

$$\phi(\theta) = \langle (B_{k+1} - B_k) + \theta (\nabla^2 F - B_{k+1}), (B_{k+1} - B_k) + \theta (\nabla^2 F - B_{k+1}) \rangle, \quad (6.5)$$

$\theta \in \mathbb{R}$, which is least at $\theta = 0$ if and only if the scalar product $\langle B_{k+1} - B_k, \nabla^2 F - B_{k+1} \rangle$ is zero. This remark implies the identity

$$\|\nabla^2 F - B_{k+1}\|_F^2 = \|\nabla^2 F - B_k\|_F^2 - \|B_{k+1} - B_k\|_F^2, \qquad (6.6)$$

which is a well-known property of least squares projection methods. Thus, if F is quadratic, the symmetric Broyden method causes the Frobenius norms of the error matrices $\nabla^2 F - B_k$, $k = 1, 2, 3, \ldots$, to decrease monotonically as the iterations proceed.

Equation (6.6) is highly relevant to the important breakthrough in convergence theory by Broyden, Dennis and Moré (1973). They find that, if ∇F is available in the unconstrained minimization of $F(x)$, $x \in \mathbb{R}^n$, then usually the sequence x_k, $k = 1, 2, 3 \ldots$, converges at a superlinear rate if the matrices B_k have the property

$$\lim_{k \to \infty} \|\nabla F(x_k + \underline{\delta}_k) - \{\nabla F(x_k) + B_k \underline{\delta}_k\}\| / \|\underline{\delta}_k\| = 0, \qquad (6.7)$$

for the choices of $\underline{\delta}_k$ considered already. The term $\nabla F(x_k) + B_k \underline{\delta}_k$ is the estimate of $\nabla F(x_k + \underline{\delta}_k)$ given by the quadratic model (6.1) in the case $g_k = \nabla F(x_k)$. Many researchers had believed previously, however, that fast convergence in practice would require B_k to be sufficiently close to $\nabla^2 F(\underline{x}_k)$. Equation (6.6) shows that $\|B_{k+1} - B_k\|_F$ tends to zero as k increases. Therefore $\|B_{k+1}\underline{\delta}_k - B_k\underline{\delta}_k\|/\|\underline{\delta}_k\|$ tends to zero too, and we let B_{k+1} be constrained by condition (6.3). Thus the condition (6.7) for superlinear convergence is satisfied by the symmetric Broyden method even if $\|\nabla^2 F - B_k\|_F$ does not become small.

Some successes of the symmetric Broyden method in minimization without derivatives are stunning. In the NEWUOA software of Powell (2006), the constraints on the parameters of the new model (6.2) are the interpolation conditions (6.4) and the symmetry condition $B_{k+1}^T = B_{k+1}$. The volume of the convex hull of the points $\underline{z}_j$, $j = 1, 2, \ldots, m$, is forced to be nonzero, in order that both $\underline{g}_{k+1}$ and B_{k+1} are defined uniquely when they provide the least value of $\|B_{k+1} - B_k\|_F$ subject to the interpolation and symmetry constraints. The test function that was used most in the development of NEWUOA has the form

$$F(\underline{x}) = \sum_{i=1}^{2n} \left\{ b_i - \sum_{j=1}^{n} \Big(S_{ij} \sin(\theta_j x_j) + C_{ij} \cos(\theta_j x_j) \Big) \right\}^2, \qquad \underline{x} \in \mathbb{R}^n, \quad (6.8)$$

which is equation (8.5) of Powell (2006). Details are given there, including the choices of the parameters b_i, S_{ij}, θ_j and C_{ij} and of a starting point $\underline{x}_1$, several choices being made randomly for each n. In each experiment, the objective function (6.8) is minimized to high accuracy and the total number of calculations of $F(\underline{x})$ is noted. The average values of these counts with $m = 2n+1$ are 931, 1809, 3159 and 6013 for $n = 20, 40, 80$ and 160, respectively. We see that these figures are roughly proportional to n, which is not very surprising if one attributes the good rate of convergence to the property $\|B_{k+1} - B_k\|_F \to 0$. On the other hand, an algorithm that constructed a careful quadratic model would require more than $n^2/2$ calculations of $F(\underline{x})$. These observations are analogous to the remark that, if $\underline{\nabla} F$ is available, if $F(\underline{x})$, $\underline{x} \in \mathbb{R}^n$, is minimized by one of the methods of Section 2, and if n is large, then it is not unusual for the required accuracy to be achieved in far fewer than n iterations.

The material of this paper leans strongly towards my own contributions to nonlinear optimization. Therefore the presentation should be regarded as a personal view of an active researcher instead of an attempt at being comprehensive. Most of the algorithms that have been addressed do not require a review, because they, with several other methods, are now studied carefully in books, such as Fletcher (1987), Nocedal and Wright (1999) and Sun and Yuan (2006). The main exception is the brief consideration of minimization without derivatives in the previous paragraph, the NEWUOA software being only five years old. An excellent survey of another part of this field is given by Kolda, Lewis and Torczon (2003). It includes some work on optimization without derivatives when there are constraints on the variables. There is a strong need in that area for new algorithms that provide high accuracy efficiently.

Acknowledgement

The author is grateful to a referee for several helpful comments.

References

C.G. Broyden, J.E. Dennis and J.J. Moré (1973), "On the local and super-linear convergence of quasi-Newton methods", *J. Inst. Math. Appl.*, Vol. 12, pp. 223–245.

T.F. Coleman and A.R. Conn (1982), "Nonlinear programming via an exact penalty function: Global analysis", *Math. Programming*, Vol. 24, pp. 137–161.

A.R. Conn, N.I.M. Gould and Ph.L. Toint (2000), *Trust-Region Methods*, MPS/SIAM Series on Optimization, SIAM (Philadelphia).

W.C. Davidon (1959), "Variable metric method for minimization", Report ANL 5990 (rev.), Argonne National Laboratory, Illinois.

A.V. Fiacco and G.P. McCormick (1968), *Nonlinear Programming: Sequential Unconstrained Minimization Techniques*, John Wiley & Sons (New York).

R. Fletcher (1985), "An ℓ_1 penalty method for nonlinear constraints", in *Numerical Optimization 1984*, eds. P.T. Boggs, R.H. Byrd and R.B. Schnabel, SIAM (Philadelphia), pp. 26–40.

R. Fletcher (1987), *Practical Methods of Optimization*, John Wiley & Sons (Chichester).

R. Fletcher and S. Leyffer (2002), "Nonlinear programming without a penalty function", *Math. Programming*, Vol. 91, pp. 239–269.

R. Fletcher and M.J.D. Powell (1963), "A rapidly convergent descent method for minimization", *Comput. J.*, Vol. 6, pp. 163–168.

R. Fletcher and C.M. Reeves (1964), "Function minimization by conjugate gradients", *Comput J.*, Vol. 7, pp. 149–154.

M.R. Hestenes (1969), "Multiplier and gradient methods", *J. Optim. Theory Appl.*, Vol. 4, pp. 303–320.

T.G. Kolda, R.M. Lewis and V. Torczon (2003), "Optimization by direct search: new perspectives on some classical and modern methods", *SIAM Review*, Vol. 45, pp. 385–482.

J.J. Moré and D.C. Sorensen (1983), "Computing a trust region step", *SIAM J. Sci. Stat. Comput.*, Vol. 4, pp. 553–572.

J. Nocedal and S.J. Wright (1999), *Numerical Optimization*, Springer (New York).

E. Polak and G. Ribière (1969), "Note sur la convergence de méthodes de directions conjuguées", *Rev. Française Informat. Recherche Opérationnelle*, 3e Année, No. 16, pp. 35–43.

M.J.D. Powell (1969), "A method for nonlinear constraints in minimization problems", in *Optimization*, ed. R. Fletcher, Academic Press (London), pp. 283–298.

M.J.D. Powell (1970), "A hybrid method for nonlinear equations", in *Numerical Methods for Nonlinear Algebraic Equations*, ed. P. Rabinowitz, Gordon and Breach (London), pp. 87–114.

M.J.D. Powell (1978), "A fast algorithm for nonlinearly constrained optimization calculations", in *Numerical Analysis, Dundee 1977, Lecture Notes in Mathematics 630*, ed. G.A. Watson, Springer-Verlag (Berlin), pp. 144–157.

M.J.D. Powell (2006), "The NEWUOA software for unconstrained optimization without derivatives", in *Large-Scale Nonlinear Optimization*, eds. G. Di Pillo and M. Roma, Springer (New York), pp. 255–297.

R.T. Rockafellar (1973), "A dual approach to solving nonlinear programming problems by unconstrained optimization", *Math. Programming*, Vol. 5, pp. 354–373.

W. Sun and Y. Yuan (2006), *Optimization Theory and Methods: Nonlinear Programming*, Springer (New York).

S.J. Wright (1997), *Primal-Dual Interior-Point Methods*, SIAM Publications (Philadelphia).

The history and development of numerical analysis in Scotland: a personal perspective

G. Alistair Watson

Division of Mathematics
University of Dundee
Dundee DD1 4HN, Scotland
gawatson@maths.dundee.ac.uk

Abstract. An account is given of the history and development of numerical analysis in Scotland. This covers, in particular, activity in Edinburgh in the first half of the 20th century, the collaboration between Edinburgh and St Andrews in the 1960s, and the role played by Dundee from the 1970s. I will give some reminiscences from my own time at both Edinburgh and Dundee.

1 Introduction

To provide a historical account of numerical analysis (or of anything else), it is necessary to decide where to begin. If numerical analysis is defined to be the study of algorithms for the problems of continuous mathematics [16], then of course it has a very long history (see, for example, [6], [13]). But "modern" numerical analysis is inextricably linked with computing machines. It is usually associated with programmable electronic computers, and is often said to have its origins in the 1947 paper by von Neumann and Goldstine [10]. The name apparently was first used around that time, and was given formal recognition in the setting up of the Institute for Numerical Analysis, located on the campus of the University of California at Los Angeles [3]. This was a section of the National Applied Mathematics Laboratories of the National Bureau of Standards headed by J. H. Curtiss, who is often given credit for the name.

Others consider modern numerical analysis to go back further than this; for example Todd [15] suggests it begins around 1936, and cites papers by Comrie and Turing. But if what is required is a systematic study of what we now think of as numerical analysis, in conjunction with the use of calculating machines, then the origin of numerical analysis in Scotland can be traced back even further, to the early years of the 20th century.

2 Edinburgh: early years

To be specific, the story begins in Edinburgh, with the arrival of E. T. Whittaker (1873–1956). From a Lancashire family, Whittaker won a Scholarship to Trinity

I would like to dedicate this article to the memory of Gene Golub and Ron Mitchell.

College, Cambridge in 1892. He graduated in Mathematics, and he then taught at Cambridge where he developed an interest in Astronomy. This led in due course to his appointment as Astronomer Royal of Ireland in 1906, and at the same time he became Professor of Astronomy at the University of Dublin. However, on the death of George Chrystal, he came to the University of Edinburgh in 1912 to fill the Chair of Mathematics. In 1913, he opened a Mathematical Laboratory in Edinburgh where students were trained to do machine calculations. Subjects to be taught included interpolation, difference formulae, determinants, linear equations, the numerical solution of algebraic equations, numerical integration, least squares and the numerical solution of differential equations. When the list of topics became known, it generated such interest that the Edinburgh Mathematical Society organized a colloquium in the summer of 1913 with the primary aim of providing instruction in aspects of numerical mathematics.

E.T. Whittaker in 1930.

Of course, the association of Edinburgh with calculating machines can be traced back further than that, at least to John Napier (1550 – 1617), who was born in Edinburgh. He was the inventor of logarithms and "Napier's Bones" (numbering rods made of ivory, used for mechanically multiplying, dividing and taking square roots and cube roots) [9].

But the arrival of Whittaker in Edinburgh saw various branches of what we now think of as numerical analysis taught in a systematic way, probably for the first time in any British University. David Gibb was one of the two lecturers on the staff in Edinburgh when Whittaker arrived, and in 1915 he published a book

[5] based on some of this material in the Edinburgh Mathematical Tracts series (edited by Whittaker). In a "Preliminary note on computation", he says:

"Each desk [in the Mathematical Laboratory] is supplied with a copy of Barlow's Tables,... a copy of Crelle's multiplication table (which gives at sight the product of any two numbers each less than 1000) and with tables giving the values of the trigonometric functions and logarithms. These may ... be supplemented by ... any of the various calculating machines now in use ... Success in computation depends partly on the proper choice of a formula and partly on a neat and methodical arrangement of the work."

Apparently, when asked by a girl student: "Sir, what is the formula for iteration?", Gibb replied "There is no formula; you just iterate."

George Robinson was also a member of Whittaker's staff in Edinburgh, and the book by Whittaker and Robinson [19], first published in 1924, is a collection of the material taught in the Laboratory. In the preface it says: "The present volume represents courses of lectures given at different times during the years 1913 – 1923 by Professor Whittaker to undergraduate and postgraduate students in the Mathematical Laboratory of the University of Edinburgh, and may be regarded as a manual of teaching and practice of the Laboratory."

A. C. Aitken (1895–1967) was born and educated in New Zealand and came to Edinburgh in 1923 to study for a PhD with Whittaker. In 1925 he was appointed to a lectureship in Actuarial Mathematics, in due course became Reader in Statistics, and in 1946 succeeded Whittaker to the Chair in Mathematics, a post he held till his retirement in 1965. According to Tee [14], Whittaker was the only person in the UK teaching determinants and matrices when Aitken arrived in Edinburgh. Aitken became interested in this area, and applied matrix algebra to numerical analysis and statistics. What we now think of as Householder matrices appeared in Aitken's book with H. W. Turnbull, Professor of Mathematics at St Andrews [17]. Aitken also published papers on polynomial equations and eigenvalue problems: the QD algorithm developed by Heinz Rutishauser in the 1950s essentially generalizes some of his results and turns them into an effective algorithm. The acceleration process which bears his name is still used today, for example in applications to the Schwarz method for domain decomposition [4], [8]).

An honours course entitled Mathematical Laboratory existed from about 1920 to 1960. The name was changed to Numerical Analysis from the academic year 1960-61, so this was the first time in Edinburgh that this name was used for an honours course. Aitken took over the teaching from Whittaker about 1946 (although he may have contributed before that time), and continued to teach it until 1961. The Mathematics Department had lots of hand calculating machines; for the benefit of the students, Aitken would appear to operate one of the machines, but would actually do the calculations in his head. The change of name may have been at the instigation of James Fulton, who took over the teaching of the course in 1961-62, as Aitken's health was giving cause for concern.

A.C. Aitken.

In 1961, the University Grants Committee had a funding round for equipment, covering all Universities in the UK. When contacted by the Committee, the Administration in Edinburgh referred this to Aitken, who did not see any requirement for a computer. Nevertheless, Edinburgh University set up a Computer Unit in 1963 with Sidney Michaelson (1925–1991) as Director. Michaelson had been appointed to a Lectureship in Mathematics at Imperial College in London in 1949, and he worked on numerical analysis throughout the 1950s and early 1960s, but his main interest was in the design and construction of digital computers. Working with K. D. Tocher, in the early 1950s he built a machine (ICCE1), using cheap war-surplus parts, which had a novel modular design. By 1957, Michaelson and Tocher were well on the way towards the construction of a valve machine, when funding was almost completely cut off: Imperial College thought that the successful development of this machine would inhibit the acquisition of a new Ferranti Mercury computer.

In any event, Michaelson moved to Edinburgh. He brought with him M. R. Osborne as Assistant Director of the Computer Unit. Osborne was born in 1934 in Australia. After a BA degree in Melbourne, he was employed from 1957 as a Scientific Officer for the Royal Australian Navy. He was posted to the UK where he spent some time in the Mathematics Division of the Admiralty Research Laboratory (ARL). He then had spells as Lecturer at the University of Reading and also at Imperial College, during which time he completed a PhD at Imperial College, supervised by Michaelson. Osborne was mainly interested at that time in finite difference methods for both ordinary and partial differential equations. He was joined in the Computer Unit by Donald Kershaw. Kershaw became interested in numerical analysis when he joined the Mathematics Division at

ARL in 1957, after two years with Vickers-Armstrong (Aircraft) in Weybridge. Following a period at ARL (where for a time he shared an office with Osborne), he moved to Edinburgh in March 1964 to teach numerical analysis. Kershaw's main interests were differential and integral equations.

Michaelson set up a landline link from the Computer Unit to the University of Manchester Atlas Computer, which came into general use in 1963. The link was largely the result of a Manchester initiative, and was accepted by Edinburgh University following protests (mainly by chemists) about the failure to acquire a computer in the 1961 funding round. When a Post Office technician came to install the landline in the Computer Unit, he was unsure how to proceed, so Michaelson drew a circuit diagram on the spot and correctly installed the connections while the technician watched. Edinburgh was allocated 15 minutes computing time per day; programmes were written in Atlas Autocode and punched on to paper tape.

The Manchester Atlas Console.

3   How I became a numerical analyst

There was no tradition of University attendance in my family, but as I progressed through school and passed various examinations, it became clear that this would be possible. My main interests were in Science and Mathematics, and I decided I wanted to study engineering, with the intention of becoming an aeronautical engineer. When I told my Science teacher this, he gave me the only piece of career advice I ever got: "Engineers are ten a penny; do pure science." Whether this was good advice or not, I will never know, but I took it and entered the

University of Edinburgh in October 1960 to take a BSc in Physics. Of course, I also studied Mathematics, and as a first year student attended lectures by Aitken on the material in his book "Determinants and Matrices" [1]. He was always keen to highlight his prodigious memory, and some time in his lectures was usually set aside for tricks and demonstrations. But, as mentioned above, his health was failing.

As an aside, I last saw Aitken when I attended a public lecture he gave (maybe in 1964) where he argued the case for a duodecimal system for currency rather than the decimal system at that time being proposed for the UK. I remember him saying (so he probably said it more than once): "They scratch, but we punch!" first making scratching motions with his 10 fingers and thumbs and then pointing to the 12 knuckles on the fingers of one hand.

After two years studying, I decided I preferred Mathematics to Physics and switched to a degree in Mathematical Sciences. In my third year, in early 1963, I could choose between a course on Numerical Analysis and one on Statistics, and having tried both for a while, decided I did not like Statistics much and continued with the Numerical Analysis option. Topics covered included rounding errors, interpolation, orthogonal expansions, Fourier and Chebyshev series, finite differences, various difference operators, numerical integration, initial value problems for ordinary differential equations, iterative methods for solving equations, the power method for eigenvalue problems, and Gaussian elimination. In addition to large books of multiplication tables, which when opened, measured about two feet by three feet, various calculating machines were available, mainly Facits and Brunsvigas, hand operated. There were also some electric calculators. The recommended book was by R. A. Buckingham [2].

Brunsviga (left) and Facit.

I took the Numerical Analysis course in both my third and fourth years. It was taught mainly by Fulton, but in 1964, he decided to make use of the expertise available in the Computer Unit and so it contained some lectures by Osborne on the numerical solution of boundary value problems in ordinary differential equations. Also, least squares problems and harmonic analysis were treated.

During my fourth year, I attended some job interviews and accepted a job with ICI in Billingham. After my graduation ceremony in Edinburgh in July 1964, I was speaking with a class mate, Alex Wight, who told me he had a summer job in the Edinburgh University Computer Unit, and suggested I go round to see the place. Of course, there was no computer there at the time, so what there was to see was not immediately obvious. But I was introduced to Osborne, and at some point he said there was a vacancy for a Demonstrator and indicated that if I was interested, I could likely get it, with the opportunity to do an MSc. The possibility of something like this had never previously crossed my mind, and of course I already had a job. However, it was an attractive idea, the salary was the same as being offered by ICI, so I thought it over, applied, and was duly offered the job. Before accepting, I wrote to tell ICI and suggested joining them later on, but they said just to apply to them again, which of course I never did.

So I started work in the Edinburgh University Computer Unit in October 1964. I attended lectures on computing and numerical methods, given as part of a postgraduate diploma course started that year, and learned to write programmes in Atlas Autocode. Part of my job included examining output when it arrived back from Manchester, and helping users to identify programming errors. Unfortunately the link was somewhat unreliable, and for a time ordinary mail was used to send programs. So it took maybe three days to find out that a comma or something had been omitted from the program, causing it to fail: of course great care was taken to avoid trivial errors!

I also embarked on research with Osborne. He had widened his interests to include Chebyshev approximation, and he asked me to look at linear Chebyshev approximation problems, and the way in which linear programming could be used, as part of an MSc project. I wrote programmes in Atlas Autocode to run on the Manchester Atlas. But I also recall on one occasion travelling through to Glasgow to run some programmes on the KDF9 there, for which an Atlas Autocode compiler had been written.

A DEC PDP/8 was installed in the Computer Unit in late 1965 and caused great excitement.

4   St Andrews

A. R. Mitchell (1921–2007) was born in Dundee. He was educated at Morgan Academy, and in 1938 went to University College Dundee, at that time a college of the University of St Andrews, to study Mathematics. Partly due to the war, student numbers were low, and he was the only Honours student in Mathematics. On graduating in 1942, he was called up for military service and went to the wartime Ministry of Aircraft Production in London, where he remained until the end of the war. His duties included the interrogation of captured Luftwaffe pilots; some years later he met one of them at a conference.

In 1946, Mitchell decided he would like to do a PhD, and returned to Dundee to see if this was possible. There was no available supervisor in Dundee, but he made contact with D. E. Rutherford in St Andrews, who agreed to act as supervisor if Mitchell would become an Assistant Lecturer for the duration of his PhD. His thesis was concerned with relaxation methods in compressible flow (Rutherford's main research interest was Lattice Theory, so the supervision must have been fairly nominal), but he developed an interest in numerical analysis, initially as a means of tackling fluid dynamics problems using Southwell's relaxation methods.

On completion of his PhD in 1950, Mitchell remained at St Andrews as a Lecturer. In 1953, J. D. Murray became his first PhD student, working on a topic in boundary layer fluid dynamics, and his second was J. Y. Thompson, who started in 1954 working on numerical aspects of fluid dynamics. J. D. Lambert was a student at St Andrews and a member of the Honours class in 1953–54 when Mitchell taught a course in numerical analysis, the first time numerical analysis had been taught in St Andrews. After graduating, Lambert worked at the Admiralty Research Laboratory (in the Fluid Dynamics Group) from 1954–1957, before having short spells at Memorial University of Newfoundland in Canada and Magee University College, Londonderry, Northern Ireland. He returned to St Andrews as a Lecturer in 1959, and became Mitchell's third PhD student, working on an idea of Mitchell's of incorporating higher derivatives into methods for ODEs. Other PhD students who came in the 1960s were Graeme Fairweather, Sandy Gourlay, Pat Keast (who had been in my class at Edinburgh), John Morris (all supervised by Mitchell), and Brian Shaw (who was supervised by Lambert).

In 1964, St Andrews acquired an IBM 1620 Model II. Apparently this was capable of solving Laplace's equation in a cube using an optimal alternating direction finite difference method with a $5 \times 5 \times 5$ mesh in 15 minutes; in a square, a 20×20 mesh could be tackled. The Computer was housed in the Observatory, over a mile from the Department of Mathematics, and hands-on access was provided for an hour each morning and afternoon, with no exceptions, even when the printer ribbon wrapped itself round the type bar, a frequent occurrence. Batch jobs could run at other times.

In 1965, Mitchell started going to evening classes in Dundee to learn Russian. He was then able to keep up with the Russian literature and was perhaps one of the first in the West to appreciate the importance of the work being done in the USSR on high order difference methods for PDEs, in particular by Samarskii, Andreyev and D'Yakonov. He met D'Yakonov at the ICM meeting in Moscow in 1966, and as a consequence, the latter visited Mitchell a few years later. A byproduct was that many others in the West became much more aware of the activity in the USSR concerning split operator techniques.

5   Collaboration between St Andrews and Edinburgh

A numerical analysis course was given in the University of Aberdeen in 1958–59 by F. W. Ponting. This was the first in Aberdeen, and was based on material from the book by D. R. Hartree [7], maybe the first book on numerical analysis to use the name. Some numerical analysis was apparently taught in the University of Glasgow in the early 1960s, using the English Electric DEUCE and then a KDF9: Glasgow was the first University in Scotland to have an electronic computer, when the DEUCE was installed in 1958. But the main centres for numerical analysis in Scotland by the mid 1960s were undoubtedly Edinburgh and St Andrews. In particular Osborne and Mitchell were keen for more interaction between their research groups and it was Osborne who initiated the idea of a conference. A detailed history of the origins of that conference are given in [18], and so a full account will not be given here. But about 25 people met in June 1965 in St Andrews for two (or maybe three) days, with Mitchell (the main organizer) and the other three members of the organizing committee (Osborne, Kershaw and Lambert) giving the main talks. I still have a folder (unfortunately now empty of its original contents) with "Symposium on Solution of DEs, St Andrews, June 1965" written on it. Although intended mainly for the two Scottish groups, a number of English based numerical analysts attended, with John Mason from Oxford probably travelling furthest.

A. R. Mitchell (left) and M. R. Osborne in Dundee in 1997

A one-day meeting on Chebyshev approximation was also organized by Osborne in Edinburgh, which was attended by the St Andrews group and others from as far afield as London. There were four speakers: Osborne and I gave talks, and also M. J. D. Powell and A. R. Curtis, both of whom had been invited up from Harwell. It was the first talk I had ever given, I was in very distinguished company, and I remember it as a rather traumatic experience.

Osborne left Edinburgh for Australia at the end of 1965, to become Director of the Computer Centre at the Australian National University, and I followed him there to start a PhD about four months later. Also in late 1965 Lambert moved from St Andrews to Aberdeen. However, in 1967 Mitchell organized a second meeting in St Andrews. This was held from 26-30 June, and called "Colloquium on the Numerical Solution of Differential Equations". There were 85 participants, 18 main speakers, each giving a 50 minute talk, and 19 shorter talks. Two of the main speakers were from overseas, but it was very much a UK event.

Incidentally when I arrived at ANU in June 1966, an IBM 360/50 had just been installed. The first thing I did was learn a new programming language, PL/1, which had just been developed by IBM. The name was originally NPL (New Progamming Language) but was changed to avoid confusion with the National Physical Laboratory. Programmes were put onto punched cards. At first the machine was only generally available during office hours. But I was able to go into the Computer Centre in the evening, switch it on, get it up and running (I recall this included mounting large disks on disk drives), put my deck of cards in a hopper, run my programme and print the output. After a few runs, I would switch everything off and go home. Of course usage grew rapidly, normal hours were extended, and this experience of "personal computing" did not last long! Nor did PL/1 at ANU, because Osborne decided after a visit to Europe that it should be replaced, and so we all started to use Fortran.

6   Dundee

D. S. Jones was appointed to the Ivory Chair of Mathematics in Queen's College, Dundee in 1965. Queen's College was part of the University of St Andrews, although as a consequence of the report of the Committee on Higher Education produced under the Chairmanship of Lord Robbins in 1963, the processes necessary to establish Dundee as a separate University were well under way.

Jones had the foresight to see numerical analysis as a growth area. He started up an MSc in Numerical Analysis and Programming, taught mainly by R. P. Pearce, the only staff member at Dundee at the time who could be considered a numerical analyst: Pearce had collaborated with Mitchell in the early 1960s (three joint papers were published in 1962 and 1963 on finite difference methods).

Queen's College separated from St Andrews to become the University of Dundee on 1st August 1967. Jones had obtained approval for a Chair of Numerical Analysis in Dundee, and also in 1967 Mitchell (then a Reader at St Andrews) was appointed. Mitchell was joined in Dundee by Lambert and Gourlay as Senior

Lecturer and Lecturer respectively, and research students and postdoctoral research fellows were attracted, the latter largely through funding from NCR (The National Cash Register Company) and the MoD (Ministry of Defence). Despite losing Mitchell and his group, St Andrews brought in replacements to maintain a presence in numerical analysis. In particular, G. M. Phillips, who worked mainly in Approximation Theory, was appointed from Southampton University in 1967. Other numerical analysts who came to St Andrews then or later on were M. A. Wolfe, J. H. McCabe and G. E. Bell.

But with the Edinburgh group diminished, the appointment of Mitchell resulted in the centre of gravity of numerical analysis in Scotland clearly shifting to Dundee. I arrived to take up an MoD Fellowship in September 1969, just missing the third conference, a "Conference on the Numerical Solution of Differential Equations", the first such meeting to be held in Dundee. It attracted 148 participants, with 8 invited speakers all from overseas, and around 45 others from outside the UK. So this was the first conference with a truly international flavour.

Among others to hold Research Fellowships in Dundee by 1970 were John Morris, Sean McKee and Nancy Nichols. Traffic was not all one way, of course. Pearce had left in 1967 for Imperial College (he was eventually to be appointed to a Chair in Meteorology in Reading), and Gourlay left in 1970 to work for IBM, although this was a good move for me, as I was appointed to the vacant Lectureship.

The academic year 1970-71 was a special one for numerical analysis in Dundee, and really put Dundee (and Scotland) on the numerical analysis map. Mitchell obtained funding from the UK Science Research Council for a Numerical Analysis Year in Dundee. The aim was to promote the theory of numerical methods and to upgrade the study of numerical analysis in British universities and technical colleges. This was done by arranging lecture courses, seminars and conferences in Dundee so that workers in the field would have the opportunity to hear about and to discuss recent research. Some 34 of the world's leading numerical analysts visited Dundee during this period, some for short periods and others for longer periods up to the full year. Of course I enjoyed being in the company of so many big names.

Five conferences were held in Dundee during that year. As well as three smaller meetings, there was a "Conference on Applications of Numerical Analysis" held from 23–26 March, 1971, with 170 participants, organized by Morris, by now a Lecturer in Dundee. I also organized a "Conference on Numerical Methods for Nonlinear Optimization" which was held from 28 June–1 July, 1971, with 198 participants. One of the main speakers at the March meeting was Lothar Collatz from Hamburg, making the first of many visits to Dundee. Another was G. H. Golub, who first came to Dundee in 1970, and who also returned many times. Indeed their enthusiasm and support for these and subsequent meetings played a large part in attracting participants from outside the UK. Among the UK invited speakers at the March meeting were K. W. Morton and M. J. D. Powell, both of whom had also been invited to the 1967 St Andrews meeting. They went on to be regular participants (not just as invited speakers), and the series benefitted greatly from the continuing presence of numerical analysts such as these.

In 1973, a "Conference on the Numerical Solution of Differential Equations" was held from July 3–6, organized by Morris who had taken over as the main conference organizer. There were 234 participants, with 20 invited speakers, and 43 submitted papers presented in parallel sessions. I edited the Proceedings (of the invited talks), and said in the Preface: "This was the 5th in a series of biennial conferences in numerical analysis, originating in St Andrews University, and held in Dundee since 1969". So this was perhaps the first explicit acknowledgement of the numbering system, with the March 1971 meeting probably interpreted as the fourth in the series.

Also in 1973, Roger Fletcher moved from Harwell to Dundee. He was (and continues to be) a leading figure in optimization, and this was a major strengthening of numerical analysis in Scotland, and of the numerical analysis group in Dundee. It also represented a significant broadening of the numerical analysis base, and this was reflected in the fact that the 1975 meeting was a "Conference on Numerical Analysis", a name which was retained. About 200 people attended the meeting from July 1-4, with 16 invited talks and 45 contributed talks.

So the biennial series was now well into its stride, and with a fairly well established pattern. It survived the departure from Dundee of Morris, who moved to Waterloo in 1975, leaving me to shoulder the main organizational load. D. F. Griffiths, who was appointed to Dundee in 1970, had shifted his interests to numerical analysis, and started to work with Mitchell, in particular on finite element methods. Mitchell's interest in finite element methods had begun in the late 1960s, apparently prompted by the arrival of Dick Wait as a PhD student, as he announced that he would like to do a PhD in that area. Griffiths organized the 1983 meeting when I was on sabbatical leave in Australia and New Zealand, and from 1985 onwards, we shared the organizational load.

The 1991 Conference celebrated Mitchell's 70th birthday, and an "A.R. Mitchell lecture" was established. The inaugural lecture was given by G. H. Golub, and he is pictured below at the 2005 Conference dinner, along with L. N. Trefethen (A. R. Mitchell lecturer in 2005 and after-dinner speaker in 2007). Golub and Collatz were both given honorary degrees by the University of Dundee.

7   The evolution of computing facilities in Dundee

Let me digress a little to say something about the tools which we as numerical analysts used, and the changes to computing facilities in Dundee, most of which I experienced. Of course, corresponding changes occurred in other places.

The first computer in Dundee was a Stantec Zebra, which was purchased (for £13,000) from College funds and private donations in 1961. Numerical results in the Mitchell and Pearce papers were obtained on this machine. In charge was A. J. Cole, who did his PhD in number theory, and had been appointed as a Lecturer in Mathematics in Dundee. He subsequently went on to establish

L.N. Trefethen (left) and G.H. Golub at the 2005 Conference dinner.

Computer Science in St Andrews (part of Computer Science is now housed in the Jack Cole building, a new building which was named after him in March 2005). The Zebra was a valve machine and input and output was on paper tape. Apparently the start-up procedure involved some banging with fists to ensure that the valves were all properly seated. There was no permanent storage and the memory consisted of 8K words kept on a magnetic drum. It was about as powerful as a 1990s pocket calculator, though a lot slower.

The first proper Computing Laboratory in Dundee was established in 1965 with J. M. Rushforth as Director and one other member of staff. Rushforth had been appointed a Lecturer in Mathematics in 1953, and took over from Cole in looking after the Stantec Zebra in 1962. (He remained as Director until his retirement in 1992.) In 1965, the Flowers report was published (A Report of a Joint Working Group on Computers for Research). In Scotland, Edinburgh was proposed as a regional centre. Replacements or upgrades for machines then being used (such as the Stantec Zebra in Dundee, Sirius machines at Strathclyde and Heriot-Watt, a KDF9 at Glasgow and an Elliott 803 in Aberdeen), were proposed. The recommendation for Dundee was an ICL machine, but Dundee did not like this proposal, and asked for and got an Elliott 4130 which was installed in 1967. This machine had 32K memory, 2 μsec access time, 48 bit word, magnetic tapes and Algol and Fortran as high level languages. Programmes were punched onto cards and run as batch jobs. It took about one hour every morning to "boot up " the machine and to carry out testing of store, card reader, half-inch magnetic tape, etc.

Punched card and card punch machine.

About 1971, the Kent Operating System provided on-line remote programming to the Elliott 4130 using BASIC. Teletypes laboriously printed at about 110 Baud (10 characters per second). A Modular One minicomputer was added as a front-end processor for remote access which by 1974 allowed dial-up modems at 2400 Baud. The Elliott 4130 was replaced in 1977 by a DEC System 10. Microcomputers arrived in 1979 (the BBC micro was about as powerful at the Elliott 4130), and around this time, email became possible to a limited range of contacts. The campus was wired for remote access to the DEC10. Mathematics acquired (black and white) VDUs (Visual Display Units), some of which could work at 4800 Baud, while others only managed 1200 Baud.

In 1984 JANET (Joint Academic Network) was created, which connected about 50 sites in the UK. Terminal access became available for logging into campus and national computers, for file transfer and for email. Colour screens on micros came in about 1985. In 1987 the DEC 10 in Dundee was replaced by a Prime 6350, and an Alliant miniSupercomputer was added the following year, the first true UNIX system.

But distributed systems had started to compete with large mainframe computers. Around 1987, members of the Department of Mathematics and Computer Science (as it was then) in Dundee acquired a large number of SUN Workstations. A SUN 3/160 file server was installed first and there was progressive expansion to SUN3s, then SUN4s. Email still required to use designated gateways: there was a convoluted system where each day's email was downloaded to the Prime. However, by the end of the 1980s, links became available to most networks in the world. Initially, the UK and the US adopted different address formats, which caused some problems, but this was eventually standardized to the US system.

Up to the mid 1980s, it was customary for secretaries to type papers and documents. The advent of document processing systems changed much of that. I remember first using nroff and troff (developed for the Unix operating system) to produce papers, before LATEX became universal in the numerical analysis community.

By 1990, things had moved a long way towards the kind of facilities which we take for granted today.

8 Postscript

My intention here has been primarily to try to give a systematic account of the development of numerical analysis in Scotland from my base year of 1913. I have chosen to take a particular route, which I judge to be the main highway. However, in sticking to this, I have inevitably passed many side roads without stopping, and I should not leave the impression that numerical analysis was only carried out in those places in Scotland so far mentioned.

For example L. F. Richardson (1881–1953), who is well known for his work on finite differences for solving differential equations, was an Englishman, did much of his work in England, but spent time in Scotland. From 1913 to 1916, he worked for the Meteorological Office as Superintendent of the Eskdalemuir Observatory, located in the south of Scotland. Later on, he spent the period from 1929–1940 as Principal of Paisley College of Technology and School of Art. He retired in 1940 so that he could concentrate on his research.

Ben Noble (1922–2006) was born near Aberdeen in Scotland. After graduating from the University of Aberdeen in radiophysics, he (like some others mentioned in this article) worked at the Admiralty Research Laboratory. After the war he obtained a master's degree from Cambridge, where he was influenced by lectures from Hartree, and stayed on at the newly established Cambridge Mathematical Laboratory. Following a spell working for the Anglo-Iranian Oil Company (now BP) and three years at the University of Keele, he spent the period from 1955 to 1962 at the Royal College of Science and Technology in Glasgow where he continued to teach while doing a DSc at the University of Aberdeen. He published two books on numerical analysis shortly after this [11], [12] in the Oliver and Boyd University Mathematical Texts Series, edited by Aitken and Rutherford. He moved to the USA in 1962, eventually succeeding J. B. Rosser as director of the Mathematics Research Center at the University of Wisconsin-Madison.

In 1964, the Royal College of Science and Technology became the University of Strathclyde. A Chair of Numerical Analysis was established in 1966 (one year before Dundee), and this was held by D. S. Butler till his retirement in 1994. D. M. Sloan, who joined the staff of Strathclyde in 1965, succeeded Butler, and was instrumental in building up numerical analysis in that institution.

176   G. Alistair Watson

In fact, there have been significant changes in the Scottish scene since the 1990s, precipitated mainly by alterations to University funding arrangements. Dundee lost Research Council funding for its Numerical Analysis MSc course, and although it continued to run for a few years for self-funded students, University pressure to rationalize courses with small numbers contributed to its closure in 1997. Those Mathematics Departments with secure income streams based on large undergraduate student numbers were well placed for growth, and other Universities in Scotland started to build up numerical analysis groups, in particular Strathclyde, as already mentioned, but also Edinburgh and Heriot-Watt.

D. B. Duncan did a PhD at Dundee and after spending some time in Canada moved to Heriot-Watt in 1986. He and Sloan were responsible for the organization of a one day meeting held at Strathclyde University in 1992 ("The Scottish Computational Mathematics Symposium"), whose stated aim was to bring together mathematicians and others who develop computer algorithms to solve mathematical problems. This continues as an annual event, normally with meetings alternating between Strathclyde and Heriot-Watt. The final Dundee biennial conference was held in 2007, ending a 42 year span of such meetings, although it is intended to continue the series from 2009 at Strathclyde. There are now as many numerical analysts at Strathclyde as mathematicians (of all kinds) in Dundee. Indeed Strathclyde currently has one of the largest and most diverse numerical analysis groups in the UK, and the reality is that there has over the last decade or so been a shift in the centre of gravity of numerical analysis in Scotland.

Acknowledgement I am grateful to many people whose memories I jogged in preparing this paper, in particular Graham Blackwood, Alex Craik, Graeme Fairweather, Philip Heywood, Pat Keast, Jack Lambert, John Morris, David Murie, Mike Osborne. The pictures of Whittaker and Aitken are courtesy of the School of Mathematics and Statistics, University of St Andrews, and the picture of the Manchester Atlas Console is from the Manchester Atlas site (http://www.chilton-computing.org.uk/acl/technology/atlas/p009.htm). The other pictures are from my collection.

References

1. A.C. Aitken, *Determinants and Matrices*, Oliver and Boyd, University Mathematical Texts, Edinburgh (1939).
2. R.A. Buckingham, *Numerical Methods*, Pitman, London (1957) (rev. 1962).
3. J.H. Curtiss, The Institute for Numerical Analysis at the National Bureau of Standards, *The American Mathematical Monthly*, **58**, (1951) 372–379.
4. M. Garbey, Acceleration of the Schwartz method for elliptic problems, *SIAM J. Sci. Comp.*, **26**, (2005) 1871–1893.
5. D. Gibb, *A course in interpolation and numerical integration for the mathematical laboratory*, Edinburgh Mathematical Tracts No 2, Bell and Sons, London (1915).
6. H.H. Goldstine, *A History of Numerical Analysis from the 16th through the 19th Century*, Springer-Verlag, New York, 1977. 14
7. D.R. Hartree, *Numerical Analysis*, Clarendon Press, Oxford, 1952.

8. F. Hulsemann, Aitken-Schwartz acceleration with auxiliary background grids, paper presented at the 2007 Dundee Biennial Numerical Analysis Conference.

9. R. Gittins, John Napier: Mathematician and Inventor of Early Calculating Devices, `http://www.computinghistorymuseum.org/teaching/papers/-biography/John%20Napier.pdf`

10. J. von Neumann and H. H. Goldstine, Numerical inverting of matrices of high order, *Bull. Amer. Math. Soc.*, **53**, (1947) 1021–1099.

11. B. Noble, *Numerical Methods 1: Iteration, Programming and Algebraic Equations*, Oliver and Boyd, University Mathematical Texts, Edinburgh (1964).

12. B. Noble, *Numercal Methods 2: Differences, Integration and Differential Equations*, Oliver and Boyd, University Mathematical Texts, Edinburgh (1964).

13. G.M. Phillips, Archimedes the Numerical Analyst, *The American Mathematical Monthly*, **88**, (1981) 165–169.

14. G.J. Tee, Alexander Craig Aitken: 1895–1967, in Proceedings of the A. C. Aitken Centenary Conference, eds L. Kavalieris, F. C. Lam, L. A. Roberts and J. Shanks, University of Otago Press, Dunedin, pp. 11–19 (1996).

15. J. Todd, Numerical analysis at the National Bureau of Standards, *SIAM Review* **17**, (1975) 361–370.

16. L.N. Trefethen, The definition of numerical analysis, *SIAM News*, **25** November 1992.

17. H.W. Turnbull and A.C. Aitken, *An Introduction to the Theory of Canonical Matrices*, Blackie and Sons, London, 1932.

18. G.A. Watson, The history of the Dundee numerical analysis conferences, *Mathematics Today*, **42**, (2006) 126–128.

19. E.T. Whittaker and G. Robinson, *The Calculus of Observations; a Treatise on Numerical Mathematics*, Blackie and Sons, London, 1924.

Remembering Philip Rabinowitz

Philip J. Davis[1] and Aviezri S. Fraenkel[2]

[1] Brown University
Providence, Rhode Island, USA
Philip_Davis@Brown.edu
[2] The Weizmann institute of Science
aviezri.fraenkel@weizmann.ac.il

The applied mathematician and numerical analyst Philip (Phil, Pinchas, Pinny) Rabinowitz was born in Philadelphia on August 14, 1926, and passed away on July 21, 2006, in Jerusalem. Philip Davis recounts reminiscences from his early scientific career; while Aviezri Fraenkel relates some of his activities at the Weizmann Institute of Science, where he began work in 1955, as well as snapshots from earlier periods.

Philip J. Davis

I had a long and fruitful friendship and collaboration with Phil (Pinny) Rabinowitz that began in the fall of 1952 at the National Bureau of Standards (NBS: now NIST) in Washington, D.C. When I began my employment there in the late summer of 1952, Phil was already there.

Phil (I never called him Pinny) grew up in Philadelphia. He got his Ph.D. degree from the University of Pennsylvania in 1951 under the supervision of Walter Gottschalk with a thesis titled *Normal Coverings and Uniform Spaces*. Of course, this topic in topology was irrelevant to the work of the bureau, and Phil was immediately pulled into numerical analysis, computation, programming, and running mathematical models of importance to members of other portions of the bureau and of the U.S. government.

At that time, the Bureau of Standards had one of the very few electronic digital computers in the world. It came on line in 1950 and was known as the SEAC (Standards Electronic Automatic Computer). Within a very short period of time Phil became an expert programmer on SEAC.

If I remember correctly, some of the features of SEAC were as follows: It had 128 memory cells, and one programmed it in what was called "the four address system". A line of code went typically as follows: take the number in cell 28, combine it with the number in cell 37 according to standard operation S, store the result in cell 6 and go to cell 18 to pick up the next instruction. Computations were in fixed-point arithmetic so that scalings had to be introduced to keep the numbers in bounds. The lines of code were first set out in pencil on standard

The permission to reprint this text was kindly provided by the American Mathematical Society. The paper appeared originally in the *Notes of the AMS*, vol. **54**, no. 11, pp. 1502–1506, December 2007.

coding sheets; these were transferred to punch cards or teletype tape, thence to magnetic wire from which they were inserted in SEAC.

In retrospect SEAC would be called a first generation computer. Though many numerical strategies (algorithms) had been worked out for a wide variety of mathematical problems in pre-electronic days, the new computers expanded the algorithmic possibilities tremendously. But it was important to work out by trial and error (and occasionally by theory) which of these strategies were optimal vis-a-vis the limitations of time, storage, money, and the difficulties inherent within the algorithm itself such as complexity, divergence, instability, ill-posedness, etc.

The 1950s were a transitional age computationally speaking. Until about 1955 or so, the electronic computers were still grinding out tables of Special Mathematical Functions and publishing them in bound volumes. Later, this was seen as largely unnecessary; special software would be incorporated into scientific computational packages and would produce values of special functions on call and as needed.

One of Phil's first publications (1954) was a *Table of Coulomb Wave Functions* done jointly with Milton Abramowitz (head of the Bureau of Standards Computation Laboratory) and Carl-Erik Fröberg, a numerical analyst from Lund, Sweden.

Shortly after I arrived in Washington, Phil worked on a project that teamed up Kenneth Cole of the National Institutes of Health and Henry Antosiewicz of NBS. Cole was a biomathematician who studied the Hodgkin-Huxley equations of impulse transmission down a nerve fiber. If I remember correctly the H-H model consisted of a system of ordinary nonlinear differential equations. Antosiewicz was an expert in that field. This very successful work was reported as "Automatic Computation of Nerve Excitation" and appeared in the Vol. 3, September 1955 issue of the *Journal of the Society for Industrial and Applied Mathematics* (SIAM).

Some incidental gossip: SIAM was founded around 1952 essentially by Ed Block who was a Ph.D. classmate of mine and who for many years was its managing director. In 1963, Alan Hodgkin and Andrew Huxley won the Nobel Prize in physiology for their work on nerve excitation, and it seems likely to me that the work of Cole, Antosiewicz, and Rabinowitz contributed a bit towards this award. Many years later, around 1988, my wife Hadassah and I met Hodgkin and his American wife socially in Cambridge, England. I told Hodgkin this NBS story, but I do not now remember what his reaction was.

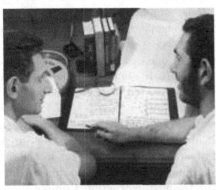

P.J. Davis (left) and P. Rabinowitz (right) laying out an algorithmic strategy. Circa 1955.

Photograph courtesy of the authors

In Washington, my friendship with Phil grew, and Hadassah and I grew to know Phil's family: his wife Terry and his children. One of his sons was born in Washington, and we were invited to the brit. There we met Phil's father and his mother. His father was a major chassidic rabbi in Philadelphia and "held court" there with many followers.

Some years later, on one of my professional trips to Philadelphia, I was able to meet Phil's sister, Margola. I believe she had or was getting a degree in philosophy from the University of Pennsylvania. She showed me around tourist Philadelphia and later we took in a summer theatre production of "Amphitryon 38" (Giradoux/S.N. Behrman) with Kitty Carlyle Hart in one of the roles. In the course of our wandering, Margola told me quite a bit about how it was growing up in a chassidic court in Philadelphia in the late 1940s. I was so amazed and intrigued by what I heard that I told her she ought to do a book of reminiscences. Perhaps she has.

In one of my first jobs at the NBS and as part of an extensive project, I was confronted with the necessity of doing some approximate integrations in the complex plane very accurately. I worked on this with Phil[3]. I thought a good strategy would be to use a very subtle and accurate scheme derived in the early 1800s by the great Carl Friedrich Gauss. Prior to 1954, the Gaussian integration rules were available only up to $n = 16$ points. The values had been calculated on desk calculators — an extremely laborious task — by Lowan, Davids, and Levenson. It was also the case that the Gaussian rules were out of favor in the days of paper-and-pencil scientific computation, as the numbers involved were helter-skelter irrational decimals, impossible to remember and difficult to enter on a keyboard without error. It was my plan to carry the computation beyond $n = 16$. I suggested to Phil that we attempt the Gaussian computation on the SEAC. He was game. I anticipated that it would be desirable to work in double-precision arithmetic to about 30 decimal places, and Phil, who was much more skillful at SEAC coding than I, agreed to write the code that would effectuate the double precision. But first I had to devise a numerical strategy. The n abscissas of the Gaussian integration rules are the roots of the Legendre polynomials of degree n. The weights corresponding to the abscissas can be obtained from the abscissas by a number of relatively simple formulas. I proposed to get the Legendre polynomials pointwise by means of the known three-term recursion relation. I would get their roots by using Newton's iterative method, starting from good approximate values. These starting values would be provided by a beautiful asymptotic formula that had been worked out in the 1930s by the Hungarian-American mathematician Gabor Szegő. I didn't know whether this strategy would work. It might fail for three or four different reasons. I was willing to try, and if it worked, good; if it didn't — well, something is always learned by failure. We could give the failure some publicity, and other mathematicians would

---

[3] P. Davis and P. Rabinowitz, "Some SEAC computations of subsonic fluid flows by Bergman's method of integral operators" (1953), in M. Z. v. Krzywoblocki, Bergman's Linear Integral Operator Method in the Theory of Compressible Fluid Flow, *Springer, Vienna, 1960.*

avoid the pitfalls and might then be able to suggest more successful strategies. I wrote the code and Phil wrote the doubleprecision part. I tried to anticipate what scaling would be necessary. I reread my code and checked it for bugs. Phil checked it for bugs. I (or Phil) punched up the code on teletype tape and checked that out. The tape was converted automatically to a wire, and the wire cartridge was inserted in the SEAC. We manually set $n = 20$, crossed our fingers, held our breath, and pushed the button to run the program.

The SEAC computed and computed and computed and computed. Our tension mounted. Finally, the computer started to output the Gaussian abscissas and weights. Numbers purporting to be such started to spew out at the teletype printer. The numbers had the right look and smell about them. We punched in $n = 24$ and again pushed the "run" button. Again, success. And ditto for even higher values of n.

The staff of the NBS computing lab declared us "Heroes of the SEAC", a title awarded in those days to programmers whose programs ran on the first try – a rare event – and for some while we had to go around wearing our "medals," which

View of SEAC, circa 1952

Photograph courtesy of the authors.

were drawn freehand in crayon on the back of used teletype paper. (The word "hero" was in parody of the practice in the Soviet Union of declaring persons "Heroes of the Soviet Union" for this and that accomplishment.)

This was the first electronic digital computation of the Gaussian integration rules. In the years since, alternative strategies have been proposed, simplified, and sharpened (by Gautschi, Golub, and others). And though all the theoretical questions that kept us guessing in 1955 have been decided positively, there are many problems as yet unsolved surrounding the Gauss idea.

Phil and I also worked together — in an experimental fashion — on the numerical solution of elliptic partial differential equations using expansions in orthogonal functions, and published a number of papers on that topic.

For Phil and me, our success and our continued interest in approximate integration led to numerous papers and to a book on the topic which, over the years, has been widely used and referenced. Our *Methods of Numerical Integration*, Academic Press, has gone through three editions. Sometime in the mid-1950s Phil decided to "make aliya" to Israel. An opportunity opened up for him at the Weizmann Instuitute of Science in Rehovot, in connection with the WEIZAC computer (1954) and the GOLEM (1964). He was hired by Chaim Pekeris who headed up the applied math group at the Weizmann Institute. Although we were now separated, our interest in producing a book on numerical integration persisted. We worked together on the book in several places; in Providence, where Phil and his family spent two semesters at Brown in 1965-6, and from February to May 1970 in Rehovot where my wife and two of our children, Ernie and Joey, spent three months. Again, in 1972, I was by myself in Rehovot for about a month, staying in the San Martin Guest House of the Weizmann Institute.

With the publication of the third edition of *Methods of Numerical Integration* in 1975, my interest in the subject slackened, though I believe that Phil published papers in the topic from time to time. He also did a book *A First Course in Numerical Analysis* with Anthony Ralston which has gone through several editions.

In between and in the years that followed, I would see Phil from time to time at conferences in different parts of the world. In 1969 we were at a conference at the University of Lancaster. The first moon landing occurred during the conference on July 20 and the sessions were suspended while — all agog — we all watched on the TV. The last day of the conference occurred on Tisha b'Av. Phil prepared to leave the conference early and return to London. I asked him why. He replied that sundown occurred earlier in London than in Lancaster and so he would be able to break his fast sooner. An example of his humor.

Aviezri S. Fraenkel [4]

Pinny (I never called him Phil) grew up in Philadelphia in a chassidic-zionist family. Since there was no Jewish day school there at the time, he studied Jewish subjects with a private tutor who came to his house for a few hours on a daily basis. While in high school, and later at the University of Pennsylvania, he attended Talmud lessons given in various synagogues in Philadelphia. He continued these studies until his deathbed.

At the university he studied medicine, but at the end of the first year he did not take a test that took place on Saturday, in order not to desecrate the sanctity of the Sabbath, so he switched to math. He got his first, second, and third degree from the University of Pennsylvania during 19461951. There was an important interlude: during 1948-9, Pinny was chosen to go to the new Servomechanism Laboratory at MIT, where he joined the Whirlwind Computer Project numerical analysis group. There he acquired his first experience in writing programs for a digital computer, interacting with people such as Alan Perlis (numerical solutions of integral equations), J. W. Carr (2-register method for floating point computations), Charles Adams (programming languages), Alex Orden and Edgar Reich (solution of linear equations). In Boston he also met Terry, whom he married shortly after getting his Ph.D. in 1951. During 1951-55 he worked at the Computation Laboratory, National Bureau of Standards, Washington, DC.

In 1954, the first digital computer in Israel was constructed under the leadership of Jerry Estrin, who was a member of the team that had just finished

---

[4] A shorter version of this part, in Hebrew, appeared recently in a Weizmann Institute publication.

constructing John von Neumann's first "stored program" computer at the Institute for Advanced Study, Princeton. Jerry later went to the Engineering Department at the University of California, Los Angeles. The initiator of WEIZAC's construction was the late Chaim L. Pekeris, head of the Applied Mathematics Department at the Weizmann Institute. The WEIZAC project was recently recognized by the Institute for Electrical and Electronics Engineers as a Milestone in the History of Computing. The unveiling of the plaque took place at the Institute on December 5, 2006. On that occasion the team members who constructed the machine received the WEIZAC Medal. Pinny and some others got it posthumously.

The front of WEIZAC

Image courtesy of the Weizmann Institute of Science.

Major operation times of WEIZAC were, addition: 50 microsecs; multiplication: 750 microsecs on the average; division: 850 microsecs. It had one of the first ferrite core memories with 4,096 words; memory access time: 10 microsecs. A unique feature of the machine was its word length: 40 bits. Input/output was via punched paper tape.

Pekeris invited Pinny to head the software development, which Pinny began in 1955, after relocating in Israel. Pinny wrote the first utility programs and built up the scientific software library, in the form of subroutines, which constituted the basic infrastructure for numerical solutions of mathematical problems. In addition he gave programming courses at various levels to many people who later became the leading programmers in Israel. In addition to Institute scientists, key personnel from government, defense, and industry participated. Pinny was the pioneer who triggered the large potential of software and high-tech industries in Israel.

Pinny taught numerical analysis at the Hebrew University, Jerusalem, and Tel Aviv and Bar Ilan Universities, in addition to the Weizmann Institute, and helped various colleges to establish computer science programs. In 1968 he received the annual prize of the Israeli Information Processing Society, the Israeli parallel of the U.S.-based Association for Computing Machinery. He traveled extensively, collaborating with mathematicians all over the continents. A conference "Numerical Integration", the core of his scientific interests, was dedicated to his sixtieth birthday. The meeting took place in Halifax, Nova Scotia, in August, 1986.

He helped the defense establishment in writing their first programs. During the tense days preceding the 6-day war, he wrote new programs and backup programs at the Institute, as fallback protection in case the defense department's main computer should become incapacitated.

Among his students were applied mathematician Nira Dyn of Tel Aviv University and computer scientist Mira Balaban of Ben Gurion University. In 1991 Mira organized an international conference on numerical analysis at Tel Aviv University, to mark Pinny's retirement. Mira is interested in artificial intelligence, especially computer music. She wrote her Ph.D. thesis on this topic, under the joint supervision of Pinny and Eli Shamir of Hebrew University. This enabled Pinny to fuse his loves for science and art.

Pinny was a passionate connoisseur of the fine arts, especially paintings, and a frequent visitor at modern art galleries. A large collection of modern paintings decorated every free inch of the walls of his home. He had a sharp eye for recognizing young talents, whose creations he purchased before they became famous, thus encouraging budding talents. As a token of thanks, some of them, such as Menashe Kadishman, dedicated some of their creations to him. He loved music ardently, especially that of Jean Sibelius.

He also encouraged and guided young mathematical talents. David Harel began concentrating on topology for his M.Sc. degree at Tel Aviv University. After one year he decided to leave his studies and become a programmer. Pinny advised him to meet Amir Pnueli. As a result, David wrote his M.Sc. degree in computer science under Amir. Both later got the Israel Prize in computer science. Amir is also a Turing Prize laureate. In June 1956 Shaula and I got married. Weeks before, Pinny secretly began hoarding the colored "holes" of the punched paper tape. When we paraded to the podium where the marriage ceremony took place, Pinny tossed the confetti on our heads. During the hot and humid summers of Rehovot, home of the Weizmann Institute, Pinny usually went abroad working with colleagues. During later years, when he reduced his travel, he purchased a house in Efrat, near Jerusalem, where his daughter lives, and the climate is cooler and drier. There he and Terry spent the summers. During winter they lived in Rehovot. Over the years, those winters became shorter and the summers got longer. During the last winter of his illness he also stayed in Efrat.

Pinny's personality reflected a harmonious fusion of Judaic values, love for the land of Israel, science, and the fine arts. May his memory be blessed.

Dedicated

to

Philip Rabinowitz

on the occasion of his sixtieth birthday, August 14, 1986

The cover page of the proceedings "Numerical Integration",
NATO Series, *Math. and Physical Sciences*, vol. 203, 1987

Image courtesy of the authors.

My early experiences with scientific computation

Philip J. Davis

Brown University
Providence, Rhode Island, USA
Philip_Davis@brown.edu

It may be a piece of vanity to write about my personal experiences as opposed to the general state of the art. But I take comfort for this limitation by the words of the great biographer Lytton Strachey:

> "Human beings are too important to be treated as mere symbols of the past. They have a value which is independent of any temporal process... and must be felt for their own sake." – *Eminent Victorians*

Hopefully, the general state of the art in the time frame I have selected will emerge from what follows here.

I attended High School in Lawrence, Massachusetts during the years 1935-1939. In our classes in "advanced" mathematics, we did some drilling with logarithms and trigonometric tables. I recall having to use the side bars that gave "proportional parts". Drilling was tedious work. In those years also, I discovered for myself what I later learned was known as the Newton forward difference interpolation formula.

Quite by accident, since my elder brother had gone to MIT, I came to own a discarded copy of David Gibbs' *A course in interpolation and numerical integration for the mathematical laboratory* London, 1915. The laboratory in question was at the University of Edinborough and run by the distinguished applied mathematician Edmund Whittaker. This laboratory was rare in academic circles. The following quote from Gibbs' book gives a vivid and amusing picture of the state of the art.

> "Each desk [in the Mathematical Laboratory at the University of Edinburgh] is equipped with a copy of *Barlow's Tables*, (square, cubes, square roots, etc.) a copy of Crelle's Tables which gives at sight the product of any two numbers less than 1000. For the neat and methodical arrangement of the work, computing paper is essential.... It will be found conducive to speed and accuracy if, instead of taking down a number one digit at a time, the computer takes it down two digits at a time."

For a graduation present from High School I asked my elder brother for a good slide rule and he gave me a Keuffel & Esser log-log slide rule – one on which fractional exponents could be done.

I attended Harvard College during the years 1939-1943, majoring in mathematics. Computation of whatever sort was not part of the curriculum. I don't believe the Department of Mathematics had an electric adding machine. One would have to go to the Physics or Astronomy Departments to find one.

There were few relevant English language books on scientific computation. One of them was E. T. Whittaker and G. Robinson's *The Calculus of Observations; a Treatise on Numerical Mathematics* (1929). Joseph Lipka's *Graphical and Mechanical Computation*, 1918, was another. I hardly need to say that there were no courses in numerical analysis (I believe this term was coined in early 1950's by George Forsythe, and the word "algorithm", though of medieval origin, had hardly been employed.)

The social status of computation among theoretical mathematicians was low; this despite the fact that famous mathematicians had worked on numerical problems related to navigation, astronomy, geodesy, tides, etc. Think of "Newton's method" of "Gauss elimination." But if one wanted to learn such material, one had to pick it up oneself.

The United States entered WW II in December, 1941. I was then in my junior year in college. Science students were draft - deferred for a while. Professors encouraged us to take a variety of science courses. The younger faculty were dropping out of teaching to join groups that worked on radar, cryptography, meteorology, operations research, feedback and control theory or the Manhattan atomic bomb project.

I received my bachelor's degree in mathematics in 1943 and in that summer, entering graduate school, I took mechanics and dynamics with Prof. Garrett Birkhoff. Shortly thereafter, since scientific talent was mobilized for the "war effort," I was alerted to and accepted a position at the National Advisory Committee for Aeronautics (NACA, the precursor of NASA), at its Langley Field, Hampton, Virginia laboratory. Somewhat later I was inducted into the United States Air Force, placed on reserve status, and sent back to the NACA with the equivalent salary of a Second Lieutenant.

As I've said, this was in the middle of WW II with the United States on a total war footing. Though there was social upset with millions of young men and many women in the armed services and such things as food and gas rationing, there was nothing in the United States that compared to the enormous physical, moral, bodily and psychological devastation experienced in Europe.

In moving from Cambridge, Massachusetts to Hampton, Virginia in early 1944, I experienced two kinds of minor shock. One was cultural, the other was scientific. The cultural shock arose from experiencing at first hand the severe prejudice and restrictions then suffered by Afro-Americans in the South of which Virginia was a part.

The scientific or organizational shock arose from this: that I (and all the young scientists who were similarly commandeered) was thrust into a well established scientific-technological environment with well seasoned old-timers, a set of problems and goals, specialized terms and ideas, and a set of preferred practices and strategies for their solution. Now sink or swim! And do more than swim: innovate if you can or if you are allowed to.

No college course can train for all the possibilities and necessities that exist in "real world" practice! In college, as a major in mathematics, I learned that the real number system was a complete Archimedean ordered field. I learned that measurable functions had little to do with physical measurements. Useless information. The noted numerical analyst and information theorist Richard Hamming wrote much later something like: "If the safety of an airplane depended on the distinction between the Riemann and the Lebesgue integrals, I would be afraid to fly."

However, from my college courses experiences, I knew something about Fourier series and complex analytic functions and these theories were quite relevant and useful. [A side remark: I did complex variable theory under David V. Widder — the "Laplace Transform Man." As reading period assignment, I studied the complex gamma function. My interest in this function continued over the years so that I wrote up as Chapter Six of the famous Abramowitz & Stegun *Handbook of Mathematical Functions*, (1964).]

The NACA was divided into a number of divisions: theoretical, full scale tunnel, compressibility, structures, etc. My job was in NACA's Aircraft Loads Division, which studied dynamic loads on the components of aircraft – often fighter planes – during a variety of flight maneuvers. I was partially a computer— working with the raw data provided by flight instrumentation, accelerometers, and pressure gauges— partially an interpreter of what I had computed. Later on, I thought of myself as an "algorithmiker", i.e., devising computational strategies.

My colleagues and I worked with slide rules, planimeters, charts, nomograms, French curves and other drawing instruments. We had various electromechanical desk-top calculators such as Marchants and Friedens. We had other computational aids: the WPA tables of special functions. We made use of these and other tables computed some years before — many in England. We had available compendia of formulas. We worked with experimental results from wind tunnel and flight data, "rules of thumb", theoretical books (e.g., the books and lectures of Ludwig Prandtl, the five volumes on aeronautics edited by William F. Durand, or the Theory of Flight by Richard von Mises). We worked with published in-house reports or reports from other laboratories that were often mimeographed or photocopied.

In those years, the word "computer" did not designate a mechanical or electronic device; it designated a person who computed. I know this at first hand because my wife (we were married early), and who had had two courses in college mathematics, was employed as a computer in the Structures Division of the NACA. It was widely believed and very likely the case that women were better and more reliable than men in carrying out computations, and in those years there were extensive employment opportunities open to them. My wife adds that the computers were treated as machines by the engineers for whom they worked: do this, do that with hardly any explanation as to what they were doing or why.

In thinking through my work at the NACA which lasted from Spring, 1944 to September, 1946, I can distinguish five major jobs that I was given to work on. The last one led to my first published paper.

1. Experimental pressure distribution on the wing profile during flight maneuvers.
2. Finding theoretically the pressure (lift) distribution over a two dimensional airfoil. (Potential flow: no compressibility, no viscosity.)
3. The inverse problem. Find the airfoil shape corresponding to a experimental pressure distribution with compressibility at higher Mach numbers.
4. Analysis of V-G diagrams (i.e., velocity-acceleration) during flight maneuvers,
5. Analysis of the failure of a flying boat tail structure under test maneuvers.

I'll now comment briefly on these jobs from the point of view of the computational procedures used.

1) A series of pressure holes over the wing profile provided the flight data. This pointwise flight data was carefully plotted and then, using French curves, "faired" to provide a continuous record. The area and the moment under the curve were then obtained by running the planimeter over the contour.

From my knowledge of the contents of Gibbs' book I felt certain that approximate integration methods applied directly to the raw data would provide equivalent accuracy, but it was done the way I just described and I had no desire to upset the computational apple cart in the middle of a war. Nonetheless, this experience and later experiences when I was employed at the National Bureau of Standards in Washington during the years of the first generation of electronic digital computers led me a to collaborate with Philip Rabinowitz on *Approximate Integration*, a book that has now gone through several editions.

2) In the years 1931-33, Theodorsen and Garrick, both employed at the NACA, had worked out a satisfactory algorithm. They mapped the exterior of the airfoil conformally onto the exterior of the unit circle (where the pressure distribution had long been known) by a rapidly convergent process involving the Joukowsky Transformation. This involved making a harmonic analysis of the airfoil contour. This analysis was accomplished by using blueprinted stencils for 24, 36, 48 point analyses. These stencils derived from the work of the German mathematician Carl Runge: *Rechenschablonen für harmonische Analyse und Synthese nach Carl Runge*, von P. Terebesi.

Runge's insight was to make use of the inherent symmetries in the sine and cosine functions when the number of points employed was highly composite. It took a computer perhaps all day to work through and check a 48 point analysis.

Here, then, was an early version of the Cooley-Tukey's FFT (the Fast Fourier Transform) which, now in chipified form, is accomplished in nanoseconds.

3) The inverse problem was: given the experimental pressure distribution from a high speed plane (high Mach numbers had already been achieved in flight) find the theoretical airfoil shape that gave rise to it under the assumptions of 1), i.e., potential flow. There was no theory behind this problem and the numerical methods I employed were essentially those of 1).

I suppose that the purpose of this investigation was to infer something about airfoil shapes that would be efficient at high Mach numbers. Despite vagueness in my mind as to what I had accomplished, I was asked to present my results to an audience of aerodynamicists and Air Force Officers. This was the first time I gave a scientific talk; just placards; no overhead projectors, no Power Point, no subsequent publication.

I've recently learned about the considerable work in the early '70's, both theoretical and algorithmic, of Paul Garabedian, David Korn et alia on the inverse problem of finding shock-free airfoil shapes.

4) The job of V-G analysis involved finding significant patterns empirically by more or less eyeballing the diagrams and doing a bit of averaging. Considering the presence of such phenomena as aerodynamic stall, his information was important in setting aerodynamically safe limits to flight maneuvers, particularly in fighter planes such as the P-40 or the P-51.

5) The Glenn Martin "Flying Boat" Mars was designed in the early '40's for the US Navy as a long range "flying dreadnought". During some initial flight-test maneuvers, the Mars, flying flat at low altitudes over Chesapeake Bay, experienced a break in its vertical tail. Why did this happen? My boss, Henry A. Pearson, a man with vast experience in aeronautics, suspected that during the testing process in which the test pilot executed the required fish-tail maneuver of oscillating the rudder, the natural frequency of the plane in flight (considered as a spring system) would be reached with a corresponding large build up of the vertical tail load.

Pearson suggested to my colleague John Boshar and me that we set up a mathematical model, use wind tunnel and various parameters from flight records, and see whether we could reproduce the build up computationally. Confining the motion to one degree of freedom (yawing motion), we set up the dynamic equation as a second order linear differential equation with the rudder deflection as the forcing function on the right hand side.

We employed the well-known method of the Duhamel Integral which solves the equation essentially as the convolution of the forcing function against a damped sinusoid, the latter requiring the eigenvalues of the differential operator. The various constants in the equation had first to be calculated as complicated combinations of aerodynamic parameters. Then, the unit impulse response – the sinusoid was calculated, and finally the convolution integral. Again we worked with a planimeter, Frieden machines and the full panoply of charts, reports, etc. from which we extracted the parameters.

Our work was successful in that it showed the possibility of serious tail overloads, and resulted in my first technical paper: *Consideration of Dynamic Loads On the Vertical Tail By The Theory Of Flat Yawing Maneuvers*. NACA, Report No.838, 1946.

The possibility and actuality of tail failure in aircraft is today still an ongoing concern. This can be learned from scanning search engine displays under this heading.

In retrospect it would have been an extremely difficult and time consuming job, in those days, to reproduce numerically the three dimensional trajectory of an airplane (pitch, yaw, and roll) corresponding to deflections of its various control surfaces. The development of the electronic digital computer received a tremendous boost from the computational necessities of the airplane and space industries.

But lest we be too proud: consider all the buildings, bridges, and planes built before electronic digital computers were available. The first supersonic airplane, the X-1, a joint project of the US Army and Air Force, the NACA, and Bell Aircraft in Buffalo, New York, flew on Oct. 14, 1947 and achieved a speed of Mach 1.06 at a time when the electronic digital computers were just getting started. The "digital wind tunnel" has not yet (2008) replaced physical experimentation, and in the opinion of some authorities, it never will.

The computers that have made flights to Mars possible have changed applied mathematics as well as our lives on Earth in ways that could not have been imagined in 1946.

Applications of Chebyshev polynomials: from theoretical kinematics to practical computations

Robert Piessens

Department of Computer Science
K.U.Leuven
Leuven, Belgium
Robert.Piessens@cs.kuleuven.be

Abstract. The Russian mathematician Pavnuty Chebyshev, born in 1821, worked on mechanical linkage design for over thirty years, which led to his work on his polynomials. The application of Chebyshev polynomials in numerical analysis starts with a paper by Lanczos in 1938. Now the computing literature abounds with papers on approximation of functions, computation of integrals and solution of differential equations, using Chebyshev polynomials. In this paper we give a survey of the role of Chebyshev polynomials in the research results of the Numerical Analysis division of the K.U.Leuven.

1 Introduction

Pavnuty Chebyshev was born in 1821 in Russia. His early education was at home where his cousin Avdotia taught him French and arithmetic. Later he would benefit from his fluency in the French language, which allowed him to communicate mathematics with the leading European mathematicians. His early research was devoted to number theory. He defended his doctoral thesis "Teoria sravneny" (Theory of congruences) in 1849. In 1850 he became extraordinary and in 1860 full professor of Mathematics at Petersburg University. This was the start of intensive research work in various fields. Besides research in probability which resulted in a generalization of the law of large numbers and a corresponding generalization of the central limit theorem of De Moivre and Laplace, he began his remarkable studies on the theory of mechanisms.

The Industrial Revolution, a period in the late 18th and early 19th century, was the golden age of mechanical linkages. Mechanical linkages are rigid links connected with joints to form a closed chain. They are designed to take an input and produce a different output, altering the motion, velocity or acceleration. Linkages are often the simplest, least expensive and most efficient mechanisms to perform complicated motions.

Design of many simple mechanisms that seem obvious today, required some of the greatest minds of the era.

Leonhard Euler (1707-1783) was one of the first mathematicians which study synthesis of linkages and James Watt (1736-1819), pioneer of the steam engine, invented the Watt linkage to support his engine's piston.

Chebyshev's interest, both in the theory of mechanisms and in the theory of approximation comes from a trip he made to France, England and Germany in 1852, where he had the chance to investigate various steam engines. From then on, he worked on mechanical linkage design for over thirty years. He studied the so-called Watt-parallelogram, a hinge mechanism employed in steam engines for transforming a rotating into a rectilinear movement. Since it is impossible to obtain strictly rectilinear movement by hinge mechanisms, Chebyshev elaborated a sound theory to reduce the deviation of the resultant movement from the rectilinear. This problem is closely related to the theory of best approximations of functions.

The paper "Théory des mécanismes connus sous le nom de parallélogrammes" (1854) was first in a series of works in this area. In this paper, Chebyshev determined the polynomials of the nth degree with leading coefficient equal to unity which deviates least from zero on the interval $[-1, 1]$. This polynomial is

$$T_n(x) = \cos(n \arccos(x))$$

and is called now the Chebyshev polynomial of degree n.

The Chebyshev linkage is a three bar mechanism that converts rotational motion to approximate straight line motion (Fig. 1).

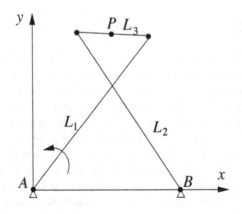

Fig. 1 The Chebyshev linkage

In Figure 1, $L_1 : L_2 : L_3 : L_4 = 5 : 5 : 2 : 4$ (where L_4 is the distance between A and B). Point P is located midway along L_3. When L_1 rotates around A, the trajectory of P is nearly a horizontal straight line. In Figure 2, the deviation of the trajectory of P from the horizontal line $y = 2.00244$ is depicted (for the size $L_1 = L_2 = 2.5$, $L_3 = 1$ and $L_2 = 2$).

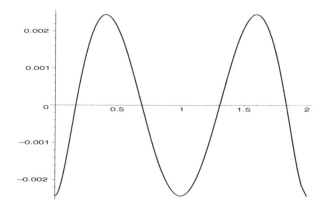

Fig. 2 The Chebyshev linkage: Deviation from the straight line.

The polynomials $T_n(x)$ form an orthogonal system on $[-1, 1]$ with respect to the weight function $(1 - x^2)^{-1/2}$.

The application of Chebyshev polynomials in numerical analysis starts with a paper of Lanczos [9] in 1938. The coming of the digital computer gave further emphasis to this development. From the middle of the 20th century, the numerical analysis literature abounds with papers on approximation of functions, computation of integrals and solution of differential equations, using Chebyshev polynomials. We mention especially the work of Lanczos [10, 11], Clenshaw [1–3], Luke [12–15], and the handbooks of Fox and Parker [6] and Rivlin [25].

Chebyshev polynomials play also an important role in network synthesis, especially for the construction of equal ripple approximations to ideal low-pass filters.

In this paper we give a survey of the use of Chebyshev polynomials in the numerical computation of integrals and integral transforms and the solution of integral equations. The survey is restricted to research results of the Numerical Analysis division of the K.U.Leuven.

2   Properties of the Chebyshev polynomials

The popularity of Chebyshev polynomials in numerical analysis is due to a lot of important but simple properties:

(i) The already mentioned property of least deviation from zero and the continuous an discrete orthogonality property.

(ii) The recurrence relation

$$T_{n+1}(x) = 2xT_n(x) - T_{n-1}(x).$$

(iii) The differential equation

$$(1 + x^2)T_n''(x) - xT_n'(x) + n^2T_n(x) = 0.$$

(iv) The difference-differential relation

$$(1 - x^2)T_n'(x) = n(T_{n-1}(x) - xT_n(x)) = \frac{n}{2}(T_{n-1}(x) - T_{n+1}(x)).$$

(v) The expression for the indefinite integral

$$\int T_n(x)dx = \frac{1}{2}\left(\frac{T_{n+1}(x)}{n+1} - \frac{T_{n-1}(x)}{n-1}\right), \quad n \geq 1.$$

(vi) The expression as a hypergeometric function

$$T_n(x) = F\left(-n, n; \frac{1}{2}; \frac{1-x}{2}\right).$$

(vii) The relatively easy formulae for constructing near least-squares approximations of a function $f(x)$:

$$f(x) \approx \sum_{k=0}^{n} {}' c_k T_k(x), \quad x \in [-1, 1],$$

$$c_k = \frac{2}{N} \sum_{l=0}^{N} {}'' f(x_l)T_l(x_k),$$

$$x_k = \cos\left(\frac{k\pi}{N}\right), \quad n \leq N,$$

where the prime denotes that the first term is taken with factor $\frac{1}{2}$, and where the double prime denotes that the first and the last term was taken with factor $\frac{1}{2}$.

The Chebyshev coefficients c_k can be evaluated using an efficient and numerically stable algorithm, based on FFT-techniques [8].

(viii) Chebyshev polynomials are members of larger families of orthogonal polynomials. (Jacobi polynomials and ultraspherical polynomials.) In many practical cases, the Chebyshev series expansion of a function is the best between all expansions into ultraspherical polynomials.

3   Inversion of the Laplace transform

The main difficulty in applying Laplace transform techniques is the determination of the original function $f(t)$ from its transform $F(p)$. In many cases, numerical methods must be used. The computation of $f(t)$ from values of $F(p)$ on the real axis is not well-posed, so that regularization is recommended. Inverting the approximation

$$F(p) \approx p^{-\alpha} \sum_{n=0}^{N} c_n T_n\left(1 - \frac{b}{p}\right) \tag{3.1}$$

yields

$$f(t) \approx \frac{t^{\alpha-1}}{\Gamma(\alpha)} \sum_{n=0}^{N} c_n \varphi_n \left(\frac{bt}{2}\right), \qquad (3.2)$$

where

$$\varphi_n(x) = {}_2F_2 \left(\begin{matrix} -n, n \\ 1/2, \alpha \end{matrix}; x\right).$$

Here, $\varphi_n(x)$ is a polynomial of degree n which satisfies the recurrence formulae [17]

$$\varphi_n(x) + (A + Bx)\varphi_{n-1}(x) + (C + Dx)\varphi_{n-2}(x) + E\varphi_{n-3}(x) = 0, \quad n = 3, 4, \ldots$$

where

$$A = -\frac{\alpha n + 3n^2 - 9n - 3\alpha + 6}{(n-2)(\alpha+n-1)},$$

$$B = \frac{4}{\alpha+n-1},$$

$$C = \frac{n(3n - 9 - \alpha) + 6}{(n-2)(\alpha+n-1)},$$

$$D = 4\frac{n-1}{(n-2)(\alpha+n-1)},$$

$$E = \frac{(n-1)(n-\alpha-2)}{(n-2)(\alpha+n-1)},$$

$$\varphi_0(x) = 1,$$

$$\varphi_1(x) = 1 - \frac{2x}{\alpha},$$

$$\varphi_2(x) = 1 - \frac{8x}{\alpha} + \frac{8x^2}{\alpha(\alpha+1)}.$$

The polynomial $\varphi_n(x)$ has n real positive zeros. This means that the interval $[0, \infty)$ can be divided into an oscillation interval, in which lie the zeros and an interval in which the polynomial increases monotonically. In the oscillation interval, $\varphi_n(x)$ oscillates with strongly increasing amplitude. In evaluating expression (3.2), this fact produce some difficulty, because, for large values of t, the errors on the coefficients c_n are multiplied by a large number, especially for large n. Regularization consists in restricting the value of N in (3.1).

4   Solution of the Abel integral equation

The Abel integral equation

$$\int_0^x \phi(x)(x-y)^{-\alpha}dy = f(x), \quad (0 < \alpha < 1), \qquad (4.1)$$

occurs in a number of engineering problems.

If $f(x)$ is differentiable, the solution of (4.1) is explicitly given by

$$\phi(x) = \frac{\sin(\alpha\pi)}{\pi}\left[\frac{f(0)}{x^{1-\alpha}} + \int_0^x \frac{f'(y)}{(x-y)^{1-\alpha}}dy\right].$$

However, this formula is not of practical value in problems where no explicit mathematical expression for $f(x)$ is known. In the case that $f(x)$ is obtainable only from measured data, Clenshaw's curve fitting method [2] can be used to construct an approximation in the form

$$f(x) \approx x^\beta \sum_{k=0}^N c_k T_k(1-2x),$$

where $\beta > -\alpha$ is essentially a free parameter, which can be used to optimize the approximation, taking into account a singular behaviour for $x \to 0$.

The approximate solution of (4.1) is now [23]

$$\phi(x) \approx \frac{x^{\alpha+\beta-1}}{\Gamma(\alpha+\beta)}\frac{\Gamma(1+\beta)}{\Gamma(1-\alpha)}\sum_{n=0}^N c_n f_n(x),$$

where

$$f_n(x) = {}_3F_2\left(\begin{array}{c}-n,n,\beta+1\\1/2,\alpha+\beta\end{array};x\right).$$

Using Fasenmyer's technique [24], a recurrence formula for the computation of $f_n(x)$ can be derived namely

$$f_n(x) + (A_n + B_n x)f_{n-1}(x) + (C_n + D_n x)f_{n-2}(x) + E_n f_{n-3}(x) = 0. \quad (4.2)$$

where

$$A_n = -\frac{1}{n-2}\left[n - 3 + \frac{(n-1)(2n-3)}{n+\alpha+\beta-1}\right]$$

$$B_n = 4\frac{n+\beta}{n+\alpha+\beta-1},$$

$$C_n = \frac{1}{n-2}\left[-1 + \frac{n-1}{n+\alpha+\beta-1}(3n-\alpha-\beta-5)\right],$$

$$D_n = -4\frac{(n-\beta-3)(n-1)}{(n+\alpha+\beta-1)(n-2)},$$

$$E_n = -\frac{(n-\alpha-\beta-2)(n-1)}{(n+\alpha+\beta-1)(n-2)}.$$

Starting values for (4.2) are

$$f_0(x) = 1,$$

$$f_1(x) = 1 - \frac{2(\beta+1)}{\alpha+\beta}x,$$

$$f_2(x) = 1 - \frac{8(\beta+1)}{\alpha+\beta}x + \frac{8(\beta+1)(\beta+2)}{(\alpha+\beta)(\alpha+\beta+1)}x^2.$$

The recurrence formula(4.2) is a difference equation of Poincaré's type. Forward recursion is numerically stable [16].

5   The computation of Laplace, Fourier and Hankel transforms

The Laplace transform of f is defined as

$$\mathcal{L}\{f\} = F(s) = \int_0^\infty e^{-sx} f(x) dx.$$

The function $f(x)$ is approximated on $[0, \infty)$ by

$$f(x) \approx (1+x)^{-\alpha} \sum_{k=0}^N a_k T_k^* \left(\frac{1}{1+x} \right).$$

where T_k^* is the shifted Chebyshev polynomial of degree k and where $\alpha > 0$ is a real parameter, which can be chosen freely, although its value affects strongly the quality of the approximation. The coefficients a_k are computed as the Chebyshev series coefficients of

$$g(z) = \left(\frac{z+1}{2} \right)^{-\alpha} f \left(\frac{1-z}{1+z} \right).$$

An approximation of $F(s)$ is now given by

$$F(s) \approx \sum_{k=0}^N a_k I_k(\alpha, s), \tag{5.1}$$

where

$$I_k(\alpha, s) = 2^{1-\alpha} e^{-s} \int_{-1}^{+1} (x+1)^{\alpha-2} e^{2s/(x+1)} T_k(x) dx.$$

Here $I_k(\alpha, s)$ satisfies the linear recurrence relation [18]

$$-(k+\alpha+1)I_{k+2} + 2(2s-k-2)I_{k+1} + 2(\alpha-3-4s)I_k$$
$$+2(2s+k-2)I_{k-1} + (k-\alpha-1)I_{k-2} = 0. \tag{5.2}$$

In (5.1) and (5.2), s may be replaced by $j\omega$, so that the formulae are applicable for the computation of Fourier integrals. Starting values for the recurrence relations and numerical stability are discussed in [20].

The Hankel transform of $f(x)$ is defined as

$$\mathcal{H}_\nu\{f\} = F_\nu(s) = \int_0^\infty x f(x) J_\nu(sx) dx,$$

where $J_\nu(x)$ is the Bessel function of the first kind and order ν. The inversion formula is, when $\nu > -\frac{1}{2}$:

$$f(x) = \mathcal{H}_\nu^{-1}\{F_\nu(s)\} = \int_0^\infty sF_\nu(s)J_\nu(sx)ds.$$

Both direct and inverse transform are integrals of the form

$$I(s) = \int_0^\infty \varphi(x)J_\nu(sx)dx,$$

which are difficult to compute numerically. However, of $\varphi(x)$ is rapidly decaying to zero, the infinite integration range may be truncated to a finite interval $[0, A]$. We have then

$$I(s) \approx A\int_0^1 \varphi(Ax)J_\nu(\omega)dx,$$

where $\omega = sA$.

Here the approximant on $[0, 1]$

$$\varphi(Ax) \approx x^\alpha \sum_{k=0}^N c_k T_k^*(x)$$

yields

$$I(s) \approx A\sum_{k=0}^N c_k M_k(\omega, \nu, \alpha),$$

where

$$M_k(\omega, \nu, \alpha) = \int_0^1 x^\alpha J_\nu(\omega x)T_k^*(x)dx.$$

These modified moments satisfy the following homogeneous, linear, nine-term recurrence relation:

$$\frac{\omega^2}{16}M_{k+4} + \left[(k+3)(k+3+2\alpha) + \alpha^2 - \nu^2 - \frac{\omega^2}{4}\right]M_{k+2}$$
$$+ [4(\nu^2 - \alpha^2) - 2(k+2)(2\alpha - 1)]M_{k+1}$$
$$- \left[2(k^2 - 4) + 6(\nu^2 - \alpha^2) - 2(2\alpha - 1) - \frac{3\omega^2}{8}\right]M_k$$
$$+ [4(\nu^2 - \alpha^2) + 2(k-2)(2\alpha - 1)]M_{k-1}$$
$$+ \left[(k-3)(k-3-2\alpha) + \left(\alpha^2 - \nu^2 - \frac{\omega^2}{4}\right)\right]M_{k-2} + \frac{\omega^2}{16}M_{k-4} = 0.$$
$$(5.3)$$

6   Solution of integral equations of the second kind using modified moments

We consider

$$\phi(x) = f(x) - \int_{-1}^{+1} k(x, y)\phi(y)dy, \tag{6.1}$$

where ϕ is the function to be determined. The kernel function k and the function f are given. We assume that $-1 \leq x \leq 1$. The use of Chebyshev polynomials for the numerical solution of such equations has been considered in [5, 6, 26]. The use of modified moments is proposed in [19]. The solution $\phi(x)$ of (6.1) is approximated by

$$p(x) = \omega(x) \sum_{k=0}^{N} c_k T_k(x), \tag{6.2}$$

where the coefficients c_k are to be determined. If it is known that $\phi(x)$ shows a singular behaviour, the singularities can be catched in the function $\omega(x)$.

Substituting (6.2) into (6.1) we have

$$\sum_{k=0}^{N} c_k[\omega(x)T_k(x) + I_k(x)] = f(x), \tag{6.3}$$

where

$$I_k(x) = \int_{-1}^{+1} k(x, y)\omega(y)T_k(y)dy.$$

Substituting at least $N + 1$ values of x into (6.3) yields a system of linear equations, the solution of which gives approximate values of the Chebyshev coefficients c_k.

In many practical cases, efficient evaluation of $I_k(x)$ is possible due to recurrence relations for modified moments. As an example, we consider Love's integral equation

$$\phi(x) = 1 \pm \frac{1}{\pi} \int_{-1}^{+1} \frac{a}{a^2 + (x - y)^2}\phi(y)dy$$

the solution of which is the field of two equal circular coaxial conducting disks, separated by a distance a and on equal or opposite potential, with zero potential at infinity. We choose $\omega(x) = 1$. The method of solution requires the evaluation of

$$I_k(x) = \int_{-1}^{+1} \frac{a}{a^2 + (x - y)^2}T_k(y)dy.$$

When a is small, the kernel function $a/(a^2 + (x - y)^2)$ shows a strongly peaked behaviour, which is a handicap for numerical integration.

The recurrence relation, however,

$$I_{k+2}(x) - 4xI_{k+1}(x) + (2 + 4a^2 + 4x^2)I_k(x) - 4xI_{k-1}(x) + I_{k-2}(x)$$
$$= (4a/(1 - k^2))[1 + (-1)^k] \tag{6.4}$$

allows efficient computation.

Starting values are

$$I_0(x) = \arctan\left(\frac{1 - x}{a}\right) + \arctan\left(\frac{1 + x}{a}\right),$$

$$I_1(x) = xI_0(x) + \frac{a}{2}\ln\frac{(1 - x)^2 + a^2}{(1 + x)^2 + a^2},$$

$$I_2(x) = 4xI_1(x) - (2a^2 + 2x^2 + 1)I_0(x) + 4a,$$

$$I_3(x) = -(4a^2 - 12x^2 + 3)I_1(x) - 8x(a^2 + x^2)I_0(x) + 16xa.$$

Forward recursion of (6.4) is not completely numerically stable, but the stability is sufficient for practical purposes.

7   An extension of Clenshaw-Curtis quadrature to oscillating and singular integrals

We want to compute an approximation to the integral

$$I = \int_{-1}^{+1} w(x)f(x)dx,$$

where $w(x)$ contains the singular or oscillating factor of the integrand and where $f(x)$ is smooth in $[-1, 1]$.

When

$$f(x) \approx \sum_{j=0}^{N} {}''c_jT_j(x),$$

then

$$I \approx \sum_{j=0}^{N} {}''c_jM_j,$$

where M_j is the modified moment

$$M_j = \int_{-1}^{+1} w(x)T_j(x)dx.$$

When $w(x) = 1$, this method is the Clenshaw-Curtis quadrature [3]. In [18] recurrence relations are given and discussed for the computation of the modified moments connected with the following weight functions:

$$w_1(x) = (1 - x)^\alpha (1 + x)^\beta,$$
$$w_2(x) = (1 - x)^\alpha (1 + x)^\beta \exp(-ax),$$
$$w_3(x) = (1 - x)^\alpha (1 + x)^\beta \ln((1 + x)/2) \exp(-ax),$$
$$w_4(x) = \exp(-ax^2),$$
$$w_5(x) = (1 - x)^\alpha (1 + x)^\beta \exp(-a(x + 1)^2),$$
$$w_6(x) = (1 - x)^\alpha (1 + x)^\beta \exp(-a/(x + 1)),$$
$$w_7(x) = (1 - x)^\alpha (1 + x)^\beta \exp(-a/x^2),$$
$$w_8(x) = (1 - x)^\alpha (1 + x)^\beta \exp(-a/(x + 1)^2),$$
$$w_9(x) = (1 - x)^\alpha (1 + x)^\beta \ln((1 + x)/2),$$
$$w_{10}(x) = (1 - x)^\alpha (1 + x)^\beta \ln((1 + x)/2) \ln((1 - x)/2),$$
$$w_{11}(x) = |x - a|^\alpha,$$
$$w_{12}(x) = |x - \alpha|^\alpha \operatorname{sign}(x - a),$$
$$w_{13}(x) = |x - \alpha|^\alpha \ln|x - a|,$$
$$w_{14}(x) = |x - \alpha|^\alpha \ln|x - a|\operatorname{sign}(x - a),$$

$$w_{15}(x) = (1 - x)^\alpha (1 + x)^\beta |x - a|^\gamma,$$
$$w_{16}(x) = (1 - x)^\alpha (1 + x)^\beta |x - a|^\gamma \ln|x - a|,$$
$$w_{17}(x) = [(x - b)^2 + a^2]^\alpha,$$
$$w_{18}(x) = (1 + x)^\alpha J_\nu(a(x + 1)/2).$$

The modified moments have also application in the numerically stable construction of Gaussian quadrature formulas [7, 21].

8   Chebyshev polynomials in QUADPACK

QUADPACK, a subroutine package for automatic integration [22] contains besides five general purpose integrators, four special purpose routines for the computation of integrals with oscillatory integrands, integrands with algebraic or logarithmic singularity and Cauchy principal value integrals. The rule evaluation component in the special purpose integration is based on Clenshaw-Curtis quadrature extension described in Section 7.

9   Conclusion

Chebyshev polynomials originate from Chebyshev's research in theoretical kinematics. Their application in numerical analysis started with Lanczos in 1938.

From the middle of the 20th century, they play an important part in the approximation and numerical differentiation and integration of smooth functions and the solution of differential equations. From 1970, the domain of applicability becomes much wider, going from the computation of integrals with singular integrand to the solution of integral equations and the inversion of integral transforms.

References

1. C.W. Clenshaw. A note on the summation of Chebyshev series. *Math. Tab. Wash.*, **9** (1955) 118–120.
2. C.W. Clenshaw. Curve fitting with a digital computer. *Comput. J.*, **2** (1960) 170–173.
3. C.W. Clenshaw. *Chebyshev series for mathematical functions*, volume 5 of *Math. Tab. Nat. Phip. Lab.* H.M. Stationary Office, London, 1962.
4. C.W. Clenshaw and A.R. Curtis. A method for numerical integration on an automatic computer. *Numer. Math.*, **2** (1960) 197–205.
5. D. Elliott. A Chebyshev series method for the numerical solution of Fredholm integral equations. *Comput. J.*, **6** (1961) 102–111.
6. L. Fox and I.B. Parker. *Chebyshev polynomials in numerical analysis*. Oxford University Press, London, 1968.
7. W. Gautschi. On the construction of Gaussian rules from modified moments. *Math. Comput.*, **24** (1970) 245–260.
8. W.M. Gentleman. Algorithm 424, Clenshaw-Curtis quadrature. *Comm. ACM*, **15** (1972) 353–355.
9. C. Lanczos. Trigonometric interpolation of empirical and analytic functions. *J. Math. Phys.*, **17** (1938) 123–199.
10. C. Lanczos. *Tables of Chebyshev polynomials*, volume 9 of *Applicable Mathhematics Series US Bureau Standards*. Government Printing Office, Washington, 1952.
11. C. Lanczos. *Applied Analysis*. Prentice-Hall, New York, 1957.
12. Y.L. Luke. *Integrals of Bessel functions*. McGraw-Hill Book Co, New York, 1962.
13. Y.L. Luke. *The special functions and their approximations, Vol 1 & 2*. Academic Press, New York, 1969.
14. Y.L. Luke. *Mathematical functions and their approximations*. Academic Press, New York, 1975.
15. Y.L. Luke. *Algorithms for the computation of mathematical functions*. Academic Press, New York, 1977.
16. J. Oliver. The numerical solution of linear recurrence relations. *Numer. Math.*, **11** (1968) 349–360.
17. R. Piessens. A new numerical method for the inversion of the Laplace transform. *J. Inst. Math. Appl.*, **10** (1972) 185–192.
18. R. Piessens. Modified Clenshaw-Curtis integration and applications to the numerical computation of integral transforms. In P. Keast and G. Fairweather, editors, *Numerical integrations: Recent developments, software and applications*, pages 35–51, Dordrecht, 1987. Reidel.
19. R. Piessens and M. Branders. Numerical solution of integral equations of mathematical physics, using Chebyshev polynomials. *J. Comput. Phys.*, **21** (1976) 178–196.

20. R. Piessens and M. Branders. Computation of Fourier transform integrals using Chebyshev series expansions. *Computing*, **32** (1984) 177–186.

21. R. Piessens, M. Chawla, and N. Jayarajan. Gaussian quadrature formulas for the numerical calculation of integrals with logarithmic singularity. *J. Comput. Phys.*, **21** (1976) 356–360.

22. R. Piessens, E. de Doncker-Kapenga, C.W. Überhuber, and D.K. Kahaner. *Quadpack: A subroutine package for automatic integration.* Springer, Berlin, 1983.

23. R. Piessens and P. Verbaeten. Numerical solution of the Abel integral equation. *BIT*, **13** (1973) 451–457.

24. E.D. Rainville. *Special functions.* MacMillan, New York, 1960.

25. T. Rivlin. *Chebyshev polynomials: From approximation theory to algebra and number theory.* Wiley, New York, 1990.

26. R.E. Scraton. The solution of integral equations in Chebyshev series. *Math. Comput.*, **23** (1969) 837–844.

Name Index

Abel, N.H., 197
Abramowitz, M., x, 80, 89, 180, 189
Adams, C., 183
Adams, J.C., iv, 36, 41, 43
Aitken, A.C., iii, ix, 1–19, 35, 44, 163–177
Akilov, G., 54, 59, 71
Albright, R., 123, 124, 137
Amari, S., 110, 136
Andreyev, A.S., 168
Anselone, P., 69
Anselone, P.M., 54, 64, 65, 67, 69
Antosiewicz, H., 180
Archimedes, 3
Atkinson, K., vi, xi, xii, 53, 69

Baker, C., 54, 69
Balaban, M., 185
Bank, R., 51
Bapat, R.B., 110, 135
Barlow, P., 163, 187
Baron, M.L., 9, 22
Bartlett, M.S., 135
Barton, D., 42, 43
Bashforth, F., iv, 36, 43
Bateman, H., 68, 70
Bauer, F.L., 9, 19
Beckman, A.O., 51
Belhumeur, P., 135
Bell, G., 106
Bell, G.E., 171
Bell, R.J.T., 15
Bellavia, S., 120, 135
Bellegarda, J.R., 131, 135
Bentley, J., 84, 90
Berman, A., 110, 111, 135
Bernkopf, M., 54, 70
Bernoulli, D., 14, 19
Berry, M.W., 110, 122, 128–130, 135, 138
Bertsekas, D., 124, 135
Bessel, F., 200
Bierens de Haan, D., 3

Birkhoff, G., 188
Blackwood, G., 176
Block, E., 180
Bogolyubov, N.N., 6, 19
Bohr, N.H.D., 47
Boser, B., 127, 135
Boshar, J., 191
Bowdler, H., 74, 89
Boyle, J.M., 79, 80, 91
Brakhage, H., 66, 68, 70
Bramble, J., 66, 67, 70
Branders, M., 199, 201, 204
Bray, T.A., 80, 91
Brenan, K.E., 42, 43
Brezinski, C., iii, xi, xii, 1, 19, 20
Bro, R., 115, 116, 125, 135, 136, 138
Browne, M., 110, 122, 128, 135
Broyden, C.G., 141, 156–159
Brunner, H., 54, 70
Buckingham, R.A., 166, 176
Bückner, H., 54, 58, 63, 66, 70
Bultheel, A., xi
Bunch, J., 51
Bunch, J.R., 83, 90
Burrage, K., 38, 43
Butcher, J.C., iv, v, 35, 36, 38, 39, 41–43
Butler, D.S., 175
Buydens, L.M.C., 130, 138
Byrne, G., 83

Caliceti, E., 19, 20
Campbell, S.L., 42, 43
Cantarella, J., 119, 135
Carathéodory, C., 10
Carr, J.W., 183
Carter, J.E., 46
Chabert, J.-L., 9, 20
Chan, T., 48, 50
Chandler, G., 63, 70
Chatelin, F., 54, 66, 70
Chawla, M., 203, 205

Chebyshev, P.L., 83, 166, 167, 193–196, 199, 201, 203
Chen, D., 109
Chen, X., 128, 135
Chervonenkis. A., 127
Cholesky, A.L., 145
Chrystal, G., 162
Chu, M.T., 122, 135
Cichocki, A., 110, 136
Clenshaw, C.W., 83, 90, 195, 198, 203, 204
Cody, W.J., 86, 87, 90
Cohen, D., 50
Cole, J., 50
Cole, K., 180
Coleman, T.F., 153, 159
Collatz, L., 9, 20, 171, 172
Comrie, L.J., 35, 44, 161
Conn, A.R., 153, 159
Cooley, J.W., 190
Cools, R., xi
Copernicus, 3
Cornelisz, P., 3
Cortes, C., 136
Couette, M.M.A., 48
Courant, R., 46
Cousin, V., 5
Couston, C., 44
Cox, J., 123, 124, 137
Cox, M.G., 80, 83, 90
Craik, A., 176
Crandall, M., 49
Cray, S., 76, 99
Crelle, A.L., 163, 187
Crouzeix, M., 10, 20
Curtis, A.R., 170, 202–204
Curtiss, C.F., 37, 43
Curtiss, J.H., 161, 176

Dahlquist, G., v, 35, 38, 39, 42–44
Daniel, J.W., 39
Dantzig, G.B., 80, 90
Davidon, W.C., ix, 141, 144, 145, 159
Davids, N., 181
Davis, Ph.J., vi, x, 179, 187

Dax, A., 117, 136
de Boor, C., 83, 90
de Doncker-Kapenga, E., 79, 91, 203, 205
de Hoog, F., 66, 70
de Jong, S., 115, 116, 135
De Moivre, A., 193
Deane, J., 40, 44
Delahaye, J.P., 1, 3, 20
Delves, L.M., 54, 70
Dennis, J.E., 157, 159
Descartes, R., 5
Dhillon, I.S., 117, 136, 137
Diele, F., 122, 135
Dines, W.H., 7
Ding, C., 136
Dixon, V., 87
Doedel, S., 52
Dongarra, J.J., viii, xi, xii, 83, 90, 93, 107
Du Croz, J., 83, 84, 90
Duff, I.S., 83, 90
Duhamel, J.M.C., 191
Duling, D., 123, 124, 137
Duncan, D.B., 176
Durand, W.F., 189
Dutka, J., 1, 20
D'Yakonov, A.M., 168

Edgerton, E.S., 137
Eickhout, B., 130, 138
Einstein, A., 110
Elliott, D., 201, 204
Enright, W.H., 42, 44, 88, 90
Estep, D., 49
Estrin, J., 183
Euler, L., iv, 9, 23, 36, 44, 194

Faber, N.K.M., 125, 136
Fairweather, G., 168, 176
Fasenmyer, M.C., 198
Fellen, B.M., 42, 44, 88, 90
Fenton, P.C., 19
Feynman, R.P., 51
Fiacco, A.V., 147, 159

Fletcher, R., ix, 74, 80, 90, 145, 146, 153, 155, 158, 159, 172
Flowers, B.H., 173
Ford, B., vii, xi, xii, 73, 75, 80, 84, 85, 90
Forder, H.G., 39
Fornberg, B., 50
Forsythe, G., 41, 188
Fossey, B., 75
Fourier, J., 166, 199, 205
Fox, L., vii, 74, 83, 195, 201, 204
Foxley, E., 75
Fröberg, C.-E., 180
Fraenkel, A.S., x, 179
Franc, V., 118, 136
Francis, J.G.F., 74, 90
Fredholm, I., 53–70
Friedman, J., 136
Frobenius, F.G., 110, 111
Fulton, J., 163, 167
Fürstenau, E., 14, 20

Galerkin, B., 56
Gales, M., 138
Garabedian, P., 191
Garbey, M., 176
Garbow, B.S., 79, 80, 91
Garrick, I.E., 190
Gauss, C.F., 58, 63, 117, 181
Gautschi, W., 182, 203, 204
Gear, C.W., 35, 41, 43, 44, 83
Gentleman, W.M., 196, 204
Germain-Bonne, B., 3, 20
Gibb, D., 162, 176, 187
Gibbons, A., 42, 44
Gibbs, J.W., 19
Giffin, M., 122, 128, 132, 133, 138
Gill, P.E., 80, 90, 114, 115, 136
Gill, S., 35, 40, 44
Giordano, A.A., 136
Gittins, R., 177
Gladwell, I., 83, 88
Goldstine, H.H., ii, 161, 176, 177
Golub, G.H., x, 19, 20, 41, 161, 171–173, 182
Gordan, M., 83

Gottschalk, W., 179
Gould, N.I.M., 154, 159
Gourlay, S., 168, 170, 171
Gragg, W.B., 41, 44
Graham, I., 66, 70
Grammaticos, B., 19, 21
Gray, H., 51
Greengard, L., 68, 70
Griffiths, D.F., 172
Grippo, L., 124, 136
Groetsch, C., 54, 70
Gu, L., 128, 135
Guillamet, D., 136
Guyon, I., 127, 135

Hackbusch, W., 54, 68, 70
Hadamard, J.S., 123, 124
Hague, S.J., 84, 90
Hairer, E., iv, 35, 39, 42, 44
Hall, G., 83
Hamada, K., 128, 132, 133, 138
Hammarling, S., 83, 90
Hamming, R., 189
Han, W., 54, 69
Hanke, M., 136
Hankel, H., 199
Hanson, R.J., 83, 90, 109, 113–115, 124, 137
Harel, D., 185
Harshman, R.A., 125, 136
Hart, K.C., 181
Hartree, D.R., 169, 175, 176
Hastie, T., 136
Haveliwala, T.H., 19, 20
Håvie, T., 10, 19, 20
Hayes, J.G., 80, 83, 90
Hayes, L., 74, 75, 82
Hazan, T., 138
He, X., 136
Henrici, P., 35, 41, 43, 44
Hespanha, J., 135
Hestenes, M.R., 149, 159
Heun, K., iv, 37, 44
Heywood, P., 176
Higham, N.J., 40, 44
Hilbert, D., 54, 56, 58, 62, 70

Hildebrand, F., 58, 70
Hindemarsh, A., 83
Hirayama, A., 13, 20
Hirose, H., 13, 20
Hirschfelder, J.O., 37, 43
Hlavač, V., 118, 136
Ho, N.-D., 110, 136
Hock, A., 74, 75
Hockney, R.W., 94, 107
Hodgkin, A., 180
Holme, H., 13, 20
Holtsmark, J., 10
Hopke, P.K., 125, 130, 136–138
Horner, W.G., 14
Hou, X.W., 137
Householder, A.S., 87, 163
Hoyer, P.O., 122, 128, 136
Hsu, F.M., 136
Hull, T.E., 42, 44, 83, 88, 90
Hulsemann, F., 163, 177
Hunt, J.C.R., 9, 20
Huxley, A., 180
Huygens, C., iii, 3–5, 20, 21
Hylleraas, E.A., 10

Ikebe, Y., 79, 80, 91
Isaacson, E., vi, 49
Ivanov, V., 54, 70

Jacobi, C.G.J., 196
Jaswon, M., 68, 70, 79, 90
Jayarajan, N., 203, 205
Jeltsch, R., v, xi, xii
Jentschura, U.D., 19, 20
Jeon, M., 138
Jesshope, C., 94, 107
Jones, D.S., 170
Joukowsky, N.Y., 190
Joyce, D.C., 1, 20
Judice, J.J., 138

Kadishman, M., 185
Kaganove, J.J., 87, 90
Kahan, W.M., 86, 90
Kahaner, D.K., 79, 91, 203, 205
Kamvar, S.D., 19, 20

Kantorovich, L.V., vi, 54, 59–61, 63,
    65, 70, 71
Karush, W., 110, 111
Keast, P., 168, 176
Keenan, M.R., 115, 117, 139
Keller, H.B., vi, xi, xii, 45–52
Keller, J., 46, 47, 50
Kepler, J., 3
Kershaw, D., 164
Khatri, C.G., 124, 136
Kim, B., 137
Kim, D., 117, 137
Kim, E., 137
Kim, H., 137
Kincaid, D., 83, 90
Klema, V.C., 79, 80, 91
Kolda, T.G., 158, 159
Kommerell, K., 5, 21
Korn, D., 191
Krasnoselskii, M., 59, 71
Kreiss, H.-O., 50
Kress, R., 54, 68, 71
Kriegman, D., 135
Krogh, F.T., 83, 90
Kronecker, L., 124
Krylov, N.M., 6, 19
Krylov, V.I., 54, 61, 63, 70
Kuhn, H.W., 110, 111
Kuki, H., 87
Kullback, S., 137
Kutta, M.W., iv, 36–40, 43, 44

Lagerstrom, P.A., 50
Lagrange, J.L., 111, 141
Lambert, J.D., 168, 170, 176
Lanczos, C., 193, 195, 203, 204
Langville, A., 110, 122–124, 128, 135,
    137
Laplace, P.S., 40, 168, 189, 193, 196
Larsson, S., 49
Laurent, P.J., 9, 21
Lawson, C.L., 83, 90, 109, 113–115,
    124, 137
Lax, P.D., 47
Lebesgue, H., 189
Lee, D.D., 122, 128, 137, 138

Lee, M., 137
Legendre, A.-M., 63
Leibler, R., 137
Leibniz, G.W., 14
Levenson, A., 181
Levin, A., 138
Lewis, R.M., 158, 159
Leyffer, S., 153, 159
Li, S.Z., 128, 135
Lighthill, M.J., 50
Lill, S., 74, 75, 82
Lin, C.J., 137
Lindberg, B., 42, 44
Linz, P., 54, 71
Lipka, J., 188
Lipstorp, D., 5
Liu, W., 122, 137
Lonseth, A., 54, 68, 71
Lorenz, J., 50
Love, E.R., 201
Lowan, A.N., 181
Lubin, A., 9
Luke, Y.L., 195, 204
Lyapunov. A.M., 49
Lyness, J.N., iv, vi, xi, xii, 23, 28, 34, 87, 90

Macconi, M., 120, 135
MacLaren, M.D., 80, 91
Maclaurin, C., 9, 21, 23
Magnus, W., 47
Mammone, R., 137
Manning, C.D., 19, 20
Mansfield, P., 76
Marchuk, G.I., 6, 21
Markov, A., 110, 111
Marquardt, D.W., ix
Marsaglia, G., 80, 91
Martin, G.L., 191
Martin, M., 137
Martin, R.S., 74, 89
Mason, J., 169
Matstoms, P., 119, 137
Maxwell, J.C., 15, 21
Mayers, D., 74
Mazer, A., 137

McCabe, J.H., 171
McCormick, G.P., 147, 159
McKee, S., 171
Mead, R., 80, 91
Meidell, B.Ø., 13, 21
Meinardus, G., 19, 21
Merson, R.H., 35, 40, 44
Messel, H., 40
Meuer, H.W., 93, 94, 106, 107
Meyer, C., 123, 124, 137
Meyer-Hermann, M., 19, 20
Michaelson, S., 164
Mignot, A., 10, 20
Milne, R.M., 5, 21
Mitchell, A.R., 161, 167–172
Mitchell, C., 7
Miyoshi, H., 99
Modha, D.M., 136
Mohamed, J., 54, 70
Moler, C.B., 79, 80, 83, 90, 91
Moore, G.E., 93, 104
Moore, R.E., 39
Moore, R.H., 64, 67, 69
Moré, J.J., 137, 154, 157, 159
Morini, B., 120, 135
Morris, J., 168–172, 176
Moulton, F.R., 36, 44
Movellan, J.R., 135
Murie, D., 176
Murray, J.D., 168
Murray, W., 80, 90, 114, 115, 136
Mysovskih, I., 63, 71

Naegelsbach, H., 21
Nagai, A., 19, 21
Nagy, J.G., 136, 137
Napier, J., 162, 177
Navara, M., 118, 136
Neave, H., 80
Nelder, J.A., 80, 91
Neumann, C., 68
Neville, E.H., 3
Newton, I., viii, 14, 142, 143
Nichols, N., 171
Noble, B., 54, 65, 71, 175, 177
Nocedal, J., 124, 137, 158, 159

Nordsieck, A., 41, 44
Novak, M., 137
Nowak, Z., 68, 70
Nyström, E.J., vi, 53, 57, 58, 63, 71
Nørdlund, N.E., 16, 21
Nørsett, S.P., 39, 42, 44

O'Beirne, T.H., 16, 17, 21
O'Hara, H., 79
Ogborn, M.E., 16, 21
Oliver, J., 199, 204
Orden, A., 183
Osborn, J., 66, 67, 70
Osborne, M.R., 164, 169–170, 176
Osinga, H.M., vi, 45, 46
Østerby, O., 4, 21

Paatero, P., 122, 123, 128, 137
Padé, H.E., iii, 18, 39
Palmer, J., 86, 90
Papageorgiou, V., 19, 21
Park, H., 137, 138
Parker, I.B., 195, 201, 204
Patterson, T.N.L., 79
Pauca, P., 110, 122, 128, 132–135, 138, 139
Pearce, R.P., 170, 171
Pearson, H.A., 191
Pearson, K., 7
Pekeris, C., 182, 184
Pentland, A.P., 138, 139
Perlis, A., 183
Perron, O., 10, 110, 111
Petzold, L.R., 42, 43
Phillips, G.M., 171, 177
Phillips, J., 61, 71
Piatek, M., 119, 135
Picken, S.M., 80, 90
Piessens, R., x, 79, 91, 193, 204
Piper, J., 122, 128, 132, 133, 138
Plemmons, R.J., viii, xi, xii, 109–111, 122, 128, 132–135, 138, 139
Pnueli, A., 185
Poincaré, H., 199
Polak, E., 138, 146, 159

Ponting, F.W., 169
Portugal, L.F., 138
Powell, M.J.D., viii, ix, xi, xii, 80, 90, 124, 138, 141, 159, 160, 170, 171
Pozrikidis, C., 68, 71
Prandtl, L., 189
Prenter, P., 61, 71
Puri, K.K., 28, 34

Qi, H., 138

Rabinowitz, P., 49
Rabinowitz, Ph., x, 179, 190
Raghavan, T.E.S., 110, 135
Ragni, S., 122, 135
Rainville, E.D., 198, 205
Ralston, A., 183
Ramadan, Z., 130, 138
Ramani, A., 19, 21
Ramath, R., 138
Rao, C.R., 124, 136
Raphson, J., 14, 143
Redivo-Zaglia, M., 1, 6, 18–20
Reeves, C.M., 146, 159
Reich, E., 183
Reid, J., 83
Reinsch, C., 74, 76, 80, 82, 83, 89, 91
Reiss, E.L., 48
Ribeca, P., 19, 20
Ribière, G., 146, 159
Richards, D.J., 15
Richardson, L.F., iii, 1–23, 175
Richter, G., 66, 71
Richtmeyer, R.D., 48
Riemann, B., 189
Rivlin, T., 195, 205
Robbins, L., 170
Robinson, G., 163, 177, 188
Rockafellar, R.T., 150, 160
Rodriguez, G., 19, 20
Rogers, D., 19
Rojas, M., 129, 138
Rokhlin, V., 68, 70
Romanus, *see* van Roomen

Romberg, W., iii, iv, 1, 5, 9, 10, 18, 19, 21
Rosen, J.B., 138
Runge, C.D.T., iv, 36–39, 43, 44, 190
Rushforth, J.M., 173
Rutherford, D.E., 168, 175
Rutishauser, H., 12, 21, 163
Ryder, B., 85, 91

Sabin, M.A., ix
Sadler, D.H., 35, 44
Saffman, P., 50
Saigey, J.F., 5, 22
Salvadori, M.G., 9, 22
Samarskii, A.A., 168
Satsuma, J., 19, 21
Saul, L.K., 128, 138
Sawaguchi, K., 14
Sayers, D.K., 83, 85, 90
Schittkowski, K., 138
Schmidt, E., 49
Schmidt, R.J., 17, 22
Schneider, C., 19, 22, 66, 71
Schonfelder, J.L., 79, 80, 91
Schwarz, H.A., 163
Sciandrone, M., 124, 136
Scott, R., 49
Scraton, R.E., 201, 205
Seatzu, S., 19, 20
Sedgwick, A.E., 42, 44, 88, 90
Seidel, P.L., 117
Sejnowski, T.J., 135
Seki Takakazu, iii
Seki, T., 13, 14, 19, 20
Seki, T.(, 13
Serra-Capizzano, S., 19, 20
Seung, H., 122, 128, 137
Sha, F., 128, 138
Shahnaz, F., 122, 138
Shaidurov, V.V., 6, 21
Shamir, E., 185
Shampine, L., 58, 69, 83
Shanks, D., iii, 16–18, 22
Shashua, A., 138
Shaw, B., 168
Shaw, N., 7

Sheppard, W.F., 5, 9, 22
Shimodaira, K., 13, 20
Shure, L., 109, 113, 115, 138
Sibelius, J., 185
Sidi, A., 1, 6, 19, 22
Simon, H.D., 93, 94, 106, 107, 136
Simpson, G.C., 8
Sloan, D.M., 175
Sloan, I., 62, 68, 71
Smith, B.T., 79, 80, 87, 91
Smith, N., 138
Smith, S.B., 22
Snel van Royen, W., 3
Snellius, see Snel van Royen, W.
Snyder, W., 138
Solomon, J., 137
Sommerfeld, A., 10
Song, X., 130, 138
Sorensen, D.C., 154, 159
Southwell, R.V., 168
Spence, A., 49, 71
Sra, S., 117, 137
Steffensen, J.F., iii, 10, 16, 21, 22
Stegun, I.A., x, 80, 89, 189
Steihaug, T., 129, 138
Stephan, E., 69, 71
Stetter, H.J., 41, 44
Stewart, G.W., 83, 90
Stiefel, E.L., 9, 22
Strachey, L., 187
Strakoš, Z., 137
Strohmaier, E., 93, 94, 106, 107
Stummel, F., 65, 71
Sun, W., 158, 160
Surzhykov, A., 19, 20
Symm, G., 68, 70, 79, 90
Szegő, G., 181

Tadmore, E., 50
Takebe, T., 13
Tapper, U., 122, 123, 128, 137
Tavener, S., 49
Taylor, B., iv, 48, 142
Taylor, G.D., 19, 21
Taylor, G.I., 50
Tee, G.J., 163, 177

Terebesi, P., 190
Theodorsen, T., 190
Thompson, J.J., 7
Thompson, J.Y., 168
Tibshirani, R., 136
Tikhonov, A.N., 19
Tocher, K.D., 164
Todd, J., 161, 177
Toint, Ph.L., 154, 159
Tokihiro, T., 19, 21
Tomasi, G., 125, 138
Toraldo, G., 137
Torczon, V., 158, 159
Trefethen, L.N., 172, 173, 177
Tucker, A.W., 110, 111
Tukey, J., 190
Turing, A., 74
Turing, A.M., ix, 161
Turk, M.A., 138, 139
Turnbull, H.W., 163, 177

Überhuber, C.W., 79, 91, 203, 205

Vainikko, G., 65–67, 71
van Benthem, M.H., 115, 117, 139
van Ceulen, L., 3, 19
Van de Graaff, R.J., 10
van der Steen, A.J., 107
van Lansbergen, P., 3
van Roomen, A., 3, 19
van Schooten, F., 5
Vandermonde, A.T., 27
Vapnik, V., 127, 135, 136, 139
Varah, J., 50
Varga, R.S., 110, 139
Verbaeten, P., 205
Vicente, L.N., 138
Vitrià, J., 136
Vogel, C.R., 136
Von Karman, T., 48
von Mises, R., 189
von Naegelsbach, H., 14
von Neumann, J., ii, iii, 161, 177,
     184

Waadeland, H., 19

Wait, D., 172
Walsh, J., 74, 75, 77, 82, 83, 88, 91
Walz, G., 1, 5, 22
Wang, H., 134, 139
Wanner, G., iv, v, xi, xii, 35, 36, 39,
     42–44
Watson, G.A., ix, xi, xii, 161, 177
Watt, J., 194
Weiss, R., 66, 70
Weniger, E.J., 1, 19, 22
Wenland, W., 69, 71, 72
Whitham, G., 50
Whittaker, E.T., ix, 15, 161–177, 187,
     188
Widder, D.V., 189
Wielandt, H., 66, 72
Wight, A., 167
Wilkinson, J.H., vii, 74, 80, 82, 83,
     87, 89, 91
Willers, I.M., 42, 43
Williams, H.C., 22
Wilson, G.V., 94, 107
Wimp, J., 1, 22
Wing, G.M., 54, 72
Wolfe, M.A., 171
Woodward, P.R., 94, 107
Wrench, Jr. J.W., 17, 22
Wright, M.H., 80, 90, 114, 115, 136
Wright, S.J., 124, 137, 149, 158–160
Wynn, P., iii, 17–18, 22

Yi, J., 122, 137
Young, A., 66, 72
Yuan, Y., 158, 160

Zahar, V.M., 42, 43
Zangwill, W., 124, 139
Zdunek, R., 110, 136
Zhang, H.J., 128, 135, 137
Zhang, P., 134, 139

Subject Index

A-stable, v, 38, 39
Abel integral equation, 197
ACM, 50
Advanced Strategic Computing, 100
air pollution, 121, 130
Algol, vii, 18, 75, 80, 82, 83, 89, 173
algorithm
    epsilon, iii
    genetic, ix
    QD, iii, 12, 163
ALS, see Alternating Least Squares
alternating least squares, 116, 123
AMD
    Opteron, 101, 102
Argonne National Laboratory, iv, 23
ASC, see Advanced Strategic Computing
Atlas
    Autocode, 165, 167
    computer, 74
    console, 165

Barlow's tables, 163, 187
Beowulf cluster, 98
Bernoulli numbers, 14
Bessel function, 200
bifurcation, vi, 48, 49
BLAS, vii, 83
blind source separation, viii
BN-stable, 39
boundary integral equations, 53
Brunsviga calculator, 73, 166

calculator
    Brunsviga, 166
    Facit, 166
    Frieden, 189
    Marchant, 189
California Institute of Technology,
    see Caltech
Caltech, vi, 48–51
Cambridge
    King's College, 7

CDC, 84
Chebyshev
    polynomial, 193–204
    series, 199
Chebyshev polynomial, 83
chemometric analysis, 122
Cholesky factorization, 145
Clenshaw-Curtis quadrature, 202, 203
cluster analysis, 131
CMFortran, 97
CMOS, see Complementary Metal-
    Oxide-Semiconductor
collocation points, 63
compact approximation, 65
Complementary Metal-Oxide-Semi-
    conductor, 95
computer graphics, vi
condition number, 24, 32
conformal map, 190
Constellation, 98
continued fraction, 18
convolution, 191
Courant institute, 46
Cray 1, 76, 101, 106
Crelle's tables, 163, 187
cubature, 23–34
curve fitting, 74, 77, 78, 83
CWI Amsterdam, v, 36, 42

Dahlquist barrier, 39, 43
Daniel-Moore conjecture, 39
data
    hyperspectral, 134
    mining, viii, 110, 136
    modeling, 112
    processing, 121
degenerate kernel, 55
determinant
    Hankel, 17
    Schwein identity, 18
    Sylvester identity, 17, 18
differential equation, iv, vi, 9, 35–44,
    47, 74, 77, 78, 83, 191

stiff, v, 37
Diophantine equation, 14
domain decomposition, 163
Duhamel integral, 191
dynamical system, vi, 45

Earth Simulator, 93, 95, 96, 99, 100
ECL, *see* emitter-coupled logic
Edinburgh Mathematical Society, 162
eigenvalue problem, 6, 49, 58, 66,
    76–78, 80, 111, 191
EISPACK, vii, 80, 82, 87
emitter-coupled logic, 95
error analysis, ii
ES, *see* Earth Simulator
ETH Zürich, v
Euclidean norm, 123, 142
Euler-Maclaurin expansion, 23, 26,
    28
extrapolation, iii, 2
    ε-algorithm, iii, 18
    Aitken's Δ^2 process, 1, 2
    quadrature, 23, 26–28
    Richardson, iii, 4, 23
    Romberg, iii
    Shanks transformation, 17

face
    detection, 127
    recognition, 128
Facit calculator, 166
factor analysis, 131
Fast Fourier Transform, 190, 196
Ferranti Mercury computer, 164
Feynman Prize, 51
FFT, *see* Fast Fourier Transform
finite
    difference, 7, 58
    element, 59
fluid dynamics, 48, 50
Fortran, vii, xi, 75, 80, 82, 83, 88,
    89, 100, 170, 173
Fourier
    coefficients, 56
    series, 19, 189
    transform, 190, 205

French curves, 190
Frieden calculator, 189
Frobenius norm, 123, 125, 157
Fujitsu, 99
functional analysis, vi

Gamma function, 189
Gaussian
    elimination, 166, 188
    quadrature, 58, 181
genetic algorithm, ix
Georgia Institute of Technology, 45
Gibbs phenomenon, 19
Gnu Scientific Library (GSL), xi
Google, ii
gradient, 120, 141, 142

Hadamard product, 123, 124
Hankel transform, 199
harmonic analysis, 190
Harvard College, 187
Harwell library, 73
Hewlett-Packard, 96, 97
High Performance Computing, 93–
    106
High Performance Fortran, 97, 100
Horner scheme, 14
HP, *see* Hewlett-Packard
HPC, *see* High Performance Com-
    puting
Hubble space telescope, 134
hypergeometric function, 196

IBM, 84, 96, 97
    BlueGene/L, 93, 95, 96, 101
    PowerXCell, 101
    Roadrunner, 95, 101
    SP system, 95
ICA, *see* Independent Component An-
    ysis
ICL, 74, 75, 84
IEEE arithmetic, 89
ill-conditioned problems, 115, 129
image processing, viii, 110, 121, 128
impulse response, 191
independent component analysis, 110

integral equation, vi, 78, 79
    Fredholm, 53, 61
    Volterra, 54
integration, vi, 23–34, 58, 74, 77, 78,
        87, 190
    Gauss, 23, 26, 32
    Gauss-Legendre, 28
    mid–point rule, 24–27
    product quadrature, 64
    Romberg, iii
    trapezoidal rule, iii, 5, 9, 25–27
Intel processor, 96, 98
interpolation, 1, 58, 63, 77, 78, 83,
        162, 166
    Newton, 14
irreducible matrix, 111
iteration, vi, 110, 113, 114, 141
    Fürstenau method, 14
    Newton-Raphson, 14

Jacobi polynomial, 196
Jacobian, vi, 111, 142, 143
Joukowsky transform, 190

Karush–Kuhn–Tucker conditions, 110,
        111, 148, 150
KDF-9, 74
Khatri-Rao product, 126
killer micro, 93
KKT, see Karush-Kuhn-Tucker
Kronecker product, 127
Krylov subspace, 68

Laboratory
    Admiralty Research —, 164
    Argonne National —, iv, 23, 80,
        82, 83, 159
    Chilton Atlas —, 75
    Harwell, 74
    Lawrence Berkley —, 93
    Lawrence Livermore National
        —, 101
    Los Alamos National —, 48, 95
    National Aerospace —, 99
    National Physical —, 74, 79, 170
    Naval Ordnance —, 17, 22

    Oak Ridge National —, 93
Lagrange multiplier, 151
Laplace
    equation, 7
    transform, 40, 189, 196, 199
latent semantic indexing, 130
Lawrence Berkeley Laboratory, 93
Lawrence Livermore National Labo-
        ratory, 101
LCP, see linear complementarity prob-
        lem
Legendre polynomial, 181
library
    elefunt, 87
    Harwell, 73
    NAG, 73–89
linear algebra, ii, 77, 78, 82, 87, 109,
        120
linear complementarity problem, 113,
        117
linear programming, 80
linear regression, 114
linkage, 193
LINPACK, vii, 83
Linux, 97, 98
LLNL, see Lawrence Livermore Na-
        tional Laboratory
Los Alamos National Laboratory, 48,
        95
Love's integral, 201
LSI, see latent semantic indexing
Lyapunov-Schmidt method, 49

M-matrix, 110
Mach number, 190
machine learning, 110
magic square, 14
Marchant calculator, 189
Markov Chain, 110
Massive Parallel Processor, 93, 95,
        96, 98
mathematical laboratory, 162, 163,
        187
Mathematical Society
    American, ii, x, 179
    Australian, 41

Edinburgh, 162
Suisse, v
Matlab, xi, 109, 113, 114, 119, 137
maximum margin hyperplane, 127
Message Passing Interface, 98
method, 127
    A-stable, v
    active–set, 114
    Adams-Bashforth, iv, 36
    Adams-Moulton, 36
    augmented Lagrangian, 141, 149
    backward difference, 42
    backward-difference, 37
    barrier, 147
    Bernoulli, 14
    BFGS, 146, 152
    Clenshaw's curve fitting, 198
    collocation, vi, 56, 61
    conjugate gradient, viii, 146
    continuation, 49
    Dahlquist, v
    degenerate kernel, 55, 59
    DFP, ix, 145, 152
    Duhamel integral, 191
    Euler, iv, 36, 37
    Fürstenau, 14
    finite element, 59, 68
    Galerkin, vi, 56
    Gauss-Seidel, 117
    Gaussian elimination, 166, 188
    Heun, iv
    Huygens, 5
    interior point, 120
    iterated projection, 62
    Lagrange multipliers, 111
    least squares, 109, 112
    Lyapunov-Schmidt, 49
    multistep, iv
    Newton, 59, 142, 151, 156, 181,
        188
    Newton-like, 121
    Newton-Raphson, 143, 152
    Nyström, vi, 55, 58, 63
    of moments, 56
    predictor-corrector, 36
    projection, 55, 60
    QL, 74
    QR, 74
    quadrature, 55
    quasi-Newton, viii, ix, 117
    Romberg, 5
    Runge-Kutta, iv, v, 36–44
    SCA, 118
    simplex, 80
    simulated annealing, ix
    steepest descent, 59, 143
    trust region, 141, 156
    variable metric, viii, 141
Microsoft Windows, 98
MIT, 183, 187
modified moment, 203
Moore's law, 93, 104
MPI, see Message Passing Interface
MPP, see Massive Parallel Processor
multicore, vii
multiplicative margin maximization,
    128

NACA, 188–192
NAG Library, 73–89
Napier's Bones, 162
NASA, 188
National Aerospace Laboratory, 99
National Bureau of Standards, x, 161
    180, 190
National Physical Laboratory, 74
NEC, 96, 100
netlib, xi
Neumann series, 68
neural learning, 122
NMF, see Nonnegative Matrix Factorization
NNLS, see nonnegative least squares
Nobel prize, 73
nomogram, 189
nonnegative
    least squares, 109–135
        Fast NNLS, 115, 116
    matrix, 110

matrix factorization, viii, 110–135

tensor factorization, 110, 112

normal equations, 115

Norwegian Institute of Technology, 10

NTF, *see* Nonnegative Tensor Factorization

Nuclear Magnetic Imaging, 76

numerical linear algebra, 59

Numerical Wind Tunnel, 99, 100

NWT, *see* Numerical Wind Tunnel

Oak Ridge National Lab., 93

Oberwolfach, 51

object characterization, 122

Octave, xi

OpenMP, 98

operations research, 110

operator theory, vi

optical character recognition, 127

optimization, viii, 74, 77, 109, 127, 141–158

order stars, 39, 42

orthogonal, iii, vi, 33, 56, 62, 143, 145–147, 149, 166, 182, 195, 196

polynomial, iii, 33, 196

outer product, 122

Padé approximant, 18, 39

PageRank, ii, 20

PARAFAC, 116

partial differential equation, 68, 110

PCA, *see* Principal Component Analysis

penalty function, 148, 153

Pentium, 98

permutation matrix, 111

Perron-Frobenius theorem, 110

polyhedron, 23

power method, 11

PowerPC, 102

PowerXCell, 101

principal component analysis, viii, 110, 130

N-mode, 116

probability matrix, 111

projected gradient, 117

pseudo-inverse, 116, 126

QD algorithm, iii, 163

QL, 87

QR algorithm, 74

QUADPACK, x, 79, 203

quadratic

programming, 80, 113, 127

termination, 144

quadrature, *see* integration

random matrix, 123

random number generator, 74, 78

Reduced Instruction Set Computer, 95

regression, 127

coefficien, 114

regularization

Kantorovich, 61

Krylov, 61

Tikhonov, 19

remote sensing, 122

Richardson's number, 8

RISC, *see* Reduced Instruction Set Computer

Royal Society

Edinburgh, ix, 16

London, 7, 8, 16

Sarah Lawrence College, 47

ScaLAPACK, vii

Schwarz method, 163

SEAC, 179

sequence transformation, 2

sequential

coordinate–wise algorithm, 118

quadratic programming, 141, 151

shared memory, 98

signal processing, viii, 110

SILLIAC, 40

simulated annealing, ix

singular value decomposition, viii, 131

slide rule, 187

SMP, *see* Symmetric MultiProcessor
sound recognition, 122
special function, 77, 79, 87, 189
spectral
    data processing, 121
    imaging sensor, 134
    unmixing, 132, 133
speech recognition, 131
spline function, ix, 59
statistics, 77, 82, 110
steepest descent, 143
supercomputing, 93, 95
superconvergence, 63
support vector machine, 127
surface fitting, 74, 77, 78, 83
Symmetric MultiProcessor, 93, 95,
    96, 98

tables
    Barlow's, 163, 187
    Crelle's, 163, 187
    logarithms, 163, 187
    trigonometric, 163, 187
    WPA, 189
Taylor series, iv, 42, 142
tensor analysis, 109, 110
text mining, 121, 129
transform
    Fast Fourier, 190, 196
    Hankel, 199
    Householder, 87
    Joukowsky, 190
    Laplace, 40, 189, 196, 199
trust region, 141, 156

UNIVAC, 47
university
    Aberdeen, 169, 173
    Auckland, v, 35
    Bari, 42
    Bath, vii
    Birmingham, 75, 79, 81
    Bristol, 45
    Brown, x, 179, 187
    Cambridge, ix, 84, 141, 162

Copenhagen, 16
Dublin, 162
Dundee, ix, 161, 167, 173
    Queen's College, 170
Edinburgh, ix, 15, 161, 166, 187
ETH Zürich, v
Genève, v
Georgia Institute of Technology,
    45
Glasgow, 15, 169, 173
Heidelberg, 10
Imperial College London, 73, 164
    171
Innsbruck, 42
Iowa, vi, 53
K.U.Leuven, iii, x, 79, 193
Keele, 175
Leeds, 74, 75, 81
Lille, iii, 1
Ludwig–Maximilians, 10
Manchester, 74, 75, 79, 81, 93,
    165, 167
Mannheim, 93
Maryland, 17
Melbourne, 164
Memorial U. Newfoundland, 168
New South Wales, iv, 23
New York, 46
Norwegian Institute of Technology, 10
Nottingham, 74, 75, 81
Otago, 15
Oxford, 74, 75, 81
Queens, 79
Reading, 164
St Andrews, ix, 163, 167, 168,
    173
Stanford, 41
Strathclyde, ix, 173, 175
    Royal College of Science and
    Technology, 175
Sydney, 40
Tennessee, viii, 93
UC Davis, 51
UCLA, 161

Ulster
  Magee college, 168
  Wake Forest, viii, 109
  Wisconsin, 175

Van de Graaff generator, 10
Vandermonde matrix, 27
Vapnik-Chervonenkis theory, 127
variable metric, 141, 152
vector computer, 93

WEIZAC project, 184
Weizmann institute of Science, 179,
        184
WPA tables, 189